Solid Surfaces, Defects and Detection

Solid Surfaces, Defects and Detection

Guest Editor
Kechen Song

Basel • Beijing • Wuhan • Barcelona • Belgrade • Novi Sad • Cluj • Manchester

Guest Editor
Kechen Song
School of Mechanical
Engineering & Automation
Northeastern University
Shenyang
China

Editorial Office
MDPI AG
Grosspeteranlage 5
4052 Basel, Switzerland

This is a reprint of the Special Issue, published open access by the journal *Coatings* (ISSN 2079-6412), freely accessible at: https://www.mdpi.com/journal/coatings/special_issues/surface_defect.

For citation purposes, cite each article independently as indicated on the article page online and as indicated below:

Lastname, A.A.; Lastname, B.B. Article Title. *Journal Name* **Year**, *Volume Number*, Page Range.

ISBN 978-3-7258-3469-3 (Hbk)
ISBN 978-3-7258-3470-9 (PDF)
https://doi.org/10.3390/books978-3-7258-3470-9

© 2025 by the authors. Articles in this book are Open Access and distributed under the Creative Commons Attribution (CC BY) license. The book as a whole is distributed by MDPI under the terms and conditions of the Creative Commons Attribution-NonCommercial-NoDerivs (CC BY-NC-ND) license (https://creativecommons.org/licenses/by-nc-nd/4.0/).

Contents

About the Editor vii

Kechen Song
Solid Surfaces, Defects and Detection
Reprinted from: *Coatings* **2024**, *14*, 1575, https://doi.org/10.3390/coatings14121575 1

Xin Wen, Jvran Shan, Yu He and Kechen Song
Steel Surface Defect Recognition: A Survey
Reprinted from: *Coatings* **2023**, *13*, 17, https://doi.org/10.3390/coatings13010017 4

Chuang Liu, Kang Su, Long Yang, Jie Li and Jingbo Guo
Detection of Complex Features of Car Body-in-White under Limited Number of Samples Using Self-Supervised Learning
Reprinted from: *Coatings* **2022**, *12*, 614, https://doi.org/10.3390/coatings12050614 34

Jiayin Liu and Jae Ho Kim
A Variable Attention Nested UNet++ Network-Based NDT X-ray Image Defect Segmentation Method
Reprinted from: *Coatings* **2022**, *12*, 634, https://doi.org/10.3390/coatings12050634 46

Jiayin Liu and Jae Ho Kim
A Novel Sub-Pixel-Shift-Based High-Resolution X-ray Flat Panel Detector
Reprinted from: *Coatings* **2022**, *12*, 921, https://doi.org/10.3390/coatings12070921 65

Yu He, Xin Wen and Jing Xu
A Semi-Supervised Inspection Approach of Textured Surface Defects under Limited Labeled Samples
Reprinted from: *Coatings* **2022**, *12*, 1707, https://doi.org/10.3390/coatings12111707 82

Chi Wan, Shuai Ma and Kechen Song
TSSTNet: A Two-Stream Swin Transformer Network for Salient Object Detection of No-Service Rail Surface Defects
Reprinted from: *Coatings* **2022**, *12*, 1730, https://doi.org/10.3390/coatings12111730 96

Han Yu, Xinyue Li, Xingjie Li, Chunyu Hou, Shangyu Liu and Huasheng Xie
A Distribution-Preserving Under-Sampling Method for Imbalance Defect Recognition in Castings
Reprinted from: *Coatings* **2022**, *12*, 1808, https://doi.org/10.3390/coatings12121808 112

Yichuan Shao, Shuo Fan, Haijing Sun, Zhenyu Tan, Ying Cai, Can Zhang and Le Zhang
Multi-Scale Lightweight Neural Network for Steel Surface Defect Detection
Reprinted from: *Coatings* **2023**, *13*, 1202, https://doi.org/10.3390/coatings13071202 122

Yao Huang, Wenzhu Tan, Liu Li and Lijuan Wu
WFRE-YOLOv8s: A New Type of Defect Detector for Steel Surfaces
Reprinted from: *Coatings* **2023**, *13*, 2011, https://doi.org/10.3390/coatings13122011 136

Xiaoe Guo, Ke Gong and Chunyue Lu
Low-Resolution Steel Surface Defects Classification Network Based on Autocorrelation Semantic Enhancement
Reprinted from: *Coatings* **2023**, *13*, 2015, https://doi.org/10.3390/coatings13122015 155

About the Editor

Kechen Song

Kechen Song received his B.S., M.S., and Ph.D. degrees from the School of Mechanical Engineering and Automation, Northeastern University, Shenyang, China, in 2009, 2011, and 2014, respectively. From 2018 to 2019, he was an academic visitor with the Department of Computer Science, Loughborough University, Loughborough, U.K. He is currently an associate professor at the School of Mechanical Engineering and Automation at Northeastern University. He is an IEEE Senior Member and has authored more than 100 referred papers including those published in IEEE TII, IEEE TIV, IEEE TCSVT, IEEE TMECH, IEEE TITS, IEEE TIE, IEEE TIM, etc. Among these, there are eleven ESI highly cited papers and one hot paper. He was included in the list of the world's top 2% of scientists published by Elsevier and Stanford University and obtained the title of Outstanding Reviewer of IEEE Transactions on Instrumentation & Measurement. He served as an associate editor of *Computers, Materials & Continua* and a guest editor of *Coatings*, *Sensors*, and *Frontiers in Robotics and AI*. He also served on the youth editorial board of the *Chinese Journal of Mechanical Engineering* (CJME). He is a regular reviewer of more than 70 journals and has reviewed over a thousand papers, including IEEE TPAMI, IEEE TNNLS, IEEE TCYB, IEEE TII, etc.

Editorial

Solid Surfaces, Defects and Detection

Kechen Song

School of Mechanical Engineering & Automation, Northeastern University, Shenyang 110819, China; songkc@me.neu.edu.cn

In the modern industrial field, particularly in steel and automobile manufacturing, detecting defects in steel surfaces is crucial to product quality and safety. Due to the complexity of the manufacturing process and the variety of defect types, traditional detection methods have struggled to meet the demand for efficient, accurate, and real-time detection. In recent years, the rapid development of deep learning, computer vision, and image processing technologies has provided new solutions for steel surface defect detection. These new technologies not only enhance detection accuracy but also significantly accelerate detection speed, promoting the advancement of industrial automation.

In their paper, *"Steel Surface Defect Recognition: A Survey"*, Wen et al. conduct a comprehensive review of the development history of steel surface defect recognition technology and systematically summarize the progression from traditional image processing to deep learning methods [1]. They compare the differences between traditional and deep learning methods in terms of detection accuracy, real-time performance, computational complexity, etc. and summarize the main challenges in the field of steel surface defects detection, which are insufficient and unbalanced data samples, real-time detection, and small object detection. Finally, they provide appropriate recommendations for resolving these issues. By summarizing and comparing a large amount of literature, their paper not only provides readers with a comprehensive view of the field, but also provides theoretical support and guidance for subsequent research. The paper has been selected as an ESI highly cited paper on multiple occasions.

In practical inspection scenarios, due to the blurriness and low resolution of steel defect images, the features learned by the network suffer from information loss, feature blurring, and confusion, which makes it difficult to recognize small defects. Guo et al. proposed using ASENet, an autocorrelation semantic enhancement network, for the classification of steel surface defects. This network extracts the basic features of an image through a backbone network, enhances them using a CS attention module, and calculates the correlation between neighboring features using an autocorrelation module. Finally, the enhanced features are combined with the basic features via residual linking to obtain self-attention features, which enhance the semantic information of the image and improve defect recognition accuracy for low-resolution images [2]. Similarly, X-ray inspection, an important nondestructive testing method, is utilized for the detection of tiny defects. Such detection relies on high-resolution imaging techniques. Liu et al. improved the resolution of X-ray imaging using sub-pixel displacement, which can clearly recognize tiny defects [3]. Combining this method with deep learning techniques facilitates more accurate steel surface defect detection. Liu et al. also designed the UNet++ network with a variable attention mechanism and incorporated a preprocessing method based on the pyramid model, which further improves the performance and visibility of the extraction of faint defects, and provides effective high dynamic range compression and defect enhancement of the 16-bit raw image, improving the detection accuracy of steel surface defects in the field of nondestructive testing [4].

Citation: Song, K. Solid Surfaces, Defects and Detection. *Coatings* **2024**, *14*, 1575. https://doi.org/10.3390/coatings14121575

Received: 3 December 2024
Accepted: 6 December 2024
Published: 17 December 2024

Copyright: © 2024 by the author. Licensee MDPI, Basel, Switzerland. This article is an open access article distributed under the terms and conditions of the Creative Commons Attribution (CC BY) license (https://creativecommons.org/licenses/by/4.0/).

Traditional methods for classifying steel surface defects improve accuracy by increasing the depth of the convolutional neural network (CNN) and the number of parameters, but this approach ignores the memory overhead of large models and the incremental gain in accuracy decreases as the number of parameters increases. To address this problem, Shao et al. proposed a multi-scale lightweight neural network (MM), which takes the fusion coding module as the core and uses a Gaussian difference pyramid to construct a multi-scale neural network, which realizes the effective extraction of features at different scales and adapts to a variety of defect types [5]. In resource-constrained industrial scenarios, this method ensures detection is accurate and efficient. Similarly, to improve the detection accuracy and efficiency, Huang et al. proposed WFRE-YOLOv8s for steel surface defect detection based on YOLOv8s [6]. They addressed the data quality imbalance and designed a new neck module, RFN, which reduces computational overhead and effectively fuses features from different scales. Finally, through the EMA attention module, valuable features are more effectively extracted, improving both the detection accuracy and speed of the model and optimizing steel surface defect detection performance.

During steel surface defect detection, the dataset often suffers from a severe class imbalance, which degrades the performance of traditional models. Yu et al. proposed a distribution-preserving undersampling method, which divides all normal samples into several subgroups through cluster analysis and recombines them into balanced datasets. This ensures that the normal samples in all balanced datasets have the same distributions as the original imbalanced dataset, effectively resolving the data imbalance [7]. On the other hand, He et al. enhanced model training using semi-supervised learning by leveraging a large amount of unlabeled data in the case of label scarcity. This approach improves defect detection accuracy for complex surface textures. Although the two papers address different data challenges, both enhance the generalizability and robustness of the model through innovative learning strategies [8].

In practice, due to the randomness of defects, their textures and shapes are often very similar to the background, making defects difficult to recognize. To address this issue, Wan et al. adopted a dual-stream Swin Transformer network architecture to process local and global features separately, further improving detection accuracy [9].

In addition to steel surface defect detection, some research has extended defect detection technology to other industrial fields. Liu et al. proposed a defect detection method applicable to automobile body-in-white, which can handle the detection of complex surface features under different working conditions [10]. This study demonstrates the potential for cross-industry application of defect detection technology, providing a reference for other industrial fields.

Conflicts of Interest: The author declares no conflict of interest.

References

1. Wen, X.; Shan, J.; He, Y.; Song, K. Steel Surface Defect Recognition: A Survey. *Coatings* **2023**, *13*, 17. [CrossRef]
2. Guo, X.; Gong, K.; Lu, C. Low-Resolution Steel Surface Defects Classification Network Based on Autocorrelation Semantic Enhancement. *Coatings* **2023**, *13*, 2015. [CrossRef]
3. Liu, J.; Kim, J.H. A Novel Sub-Pixel-Shift-Based High-Resolution X-Ray Flat Panel Detector. *Coatings* **2022**, *12*, 921. [CrossRef]
4. Liu, J.; Kim, J.H. A Variable Attention Nested UNet++ Network-Based NDT X-Ray Image Defect Segmentation Method. *Coatings* **2022**, *12*, 634. [CrossRef]
5. Shao, Y.; Fan, S.; Sun, H.; Tan, Z.; Cai, Y.; Zhang, C.; Zhang, L. Multi-Scale Lightweight Neural Network for Steel Surface Defect Detection. *Coatings* **2023**, *13*, 1202. [CrossRef]
6. Huang, Y.; Tan, W.; Li, L.; Wu, L. WFRE-YOLOv8s: A New Type of Defect Detector for Steel Surfaces. *Coatings* **2023**, *13*, 2011. [CrossRef]
7. Yu, H.; Li, X.; Li, X.; Hou, C.; Liu, S.; Xie, H. A Distribution-Preserving Under-Sampling Method for Imbalance Defect Recognition in Castings. *Coatings* **2022**, *12*, 1808. [CrossRef]
8. He, Y.; Wen, X.; Xu, J. A Semi-Supervised Inspection Approach of Textured Surface Defects Under Limited Labeled Samples. *Coatings* **2022**, *12*, 1707. [CrossRef]

9. Wan, C.; Ma, S.; Song, K. TSSTNet: A Two-Stream Swin Transformer Network for Salient Object Detection of No-Service Rail Surface Defects. *Coatings* **2022**, *12*, 1730. [CrossRef]
10. Liu, C.; Su, K.; Yang, L.; Li, J.; Guo, J. Detection of Complex Features of Car Body-in-White Under Limited Number of Samples Using Self-Supervised Learning. *Coatings* **2022**, *12*, 614. [CrossRef]

Disclaimer/Publisher's Note: The statements, opinions and data contained in all publications are solely those of the individual author(s) and contributor(s) and not of MDPI and/or the editor(s). MDPI and/or the editor(s) disclaim responsibility for any injury to people or property resulting from any ideas, methods, instructions or products referred to in the content.

Review

Steel Surface Defect Recognition: A Survey

Xin Wen [1], Jvran Shan [1], Yu He [1] and Kechen Song [2,*]

1 School of Software, Shenyang University of Technology, Shenyang 110870, China
2 School of Mechanical Engineering & Automation, Northeastern University, Shenyang 110819, China
* Correspondence: songkc@me.neu.edu.cn

Abstract: Steel surface defect recognition is an important part of industrial product surface defect detection, which has attracted more and more attention in recent years. In the development of steel surface defect recognition technology, there has been a development process from manual detection to automatic detection based on the traditional machine learning algorithm, and subsequently to automatic detection based on the deep learning algorithm. In this paper, we discuss the key hardware of steel surface defect detection systems and offer suggestions for related options; second, we present a literature review of the algorithms related to steel surface defect recognition, which includes traditional machine learning algorithms based on texture features and shape features as well as supervised, unsupervised, and weakly supervised deep learning algorithms (Incomplete supervision, inexact supervision, imprecise supervision). In addition, some common datasets and algorithm performance evaluation metrics in the field of steel surface defect recognition are summarized. Finally, we discuss the challenges of the current steel surface defect recognition algorithms and the corresponding solutions, and our future work focus is explained.

Keywords: steel; automated defect detection; deep learning; survey

1. Introduction

Steel is one of the most common metal materials in our daily life, where its uses are numerous, and steel is the perfect material in many fields. Steel is widely used in civil engineering infrastructure, aerospace, shipbuilding, automotive, machinery manufacturing, and the manufacture of various household tools. According to the World Steel Association, the global steel demand is expected to continue to rise by 0.4% in 2022, with annual production reaching 1840.2 billion tons, more than all the other metals combined [1]. Today, steel is a key material for manufacturing, infrastructure, and other industries. Since the quality of steel will directly affect the quality of manufactured products and infrastructure construction, it is particularly important to control the quality of the steel produced, as this is the first guarantee of qualified products.

For the recognition of steel surface defects, it mainly includes three functions: detection, classification, and the location of defects. The detection of defects is to determine whether the inspection object contains defects; the classification of defects is, as the name implies, the classification of defect categories; and the location of defects is to determine the location of defects in the inspection object. The result of this defect location will be different for different algorithms; most algorithms display a rough anchor frame to locate the defect, while there are more accurate methods that can directly describe the shape of the defect down to the pixel level.

However, the recognition of defects on the steel surface is more difficult, and the main reasons why it is difficult to identify are as follows. First, there is an inter-class similarity and intra-class diversity of defects on the steel surface [2]. Second, there are many types of steel surface defects and some of these defects may overlap, while most classification tasks can only find the defects with the highest confidence level in the defect category, resulting in imprecise classification results [3]. In the actual production environment,

Citation: Wen, X.; Shan, J.; He, Y.; Song, K. Steel Surface Defect Recognition: A Survey. *Coatings* **2023**, *13*, 17. https://doi.org/10.3390/coatings13010017

Academic Editor: Paolo Castaldo

Received: 17 October 2022
Revised: 2 December 2022
Accepted: 4 December 2022
Published: 22 December 2022

Copyright: © 2022 by the authors. Licensee MDPI, Basel, Switzerland. This article is an open access article distributed under the terms and conditions of the Creative Commons Attribution (CC BY) license (https://creativecommons.org/licenses/by/4.0/).

it is very difficult to obtain high quality datasets for training machine learning related algorithms because defects in steel production are originally small probability events, and it is very difficult to obtain many samples with various types of defects; the labeling of the data is also costly and labor intensive [4–7]. Since the classification of defect categories is based on human subjective classification and there is still no strict classification standard, these lead to difficulties in the progress of classification work. In addition, in the actual production environment, the interference of the environment is serious such as uneven lighting, light reflection, noise, motion blur, the existence of many false defects, and a low contrast of background clutter, resulting in the recognition effect being often less than ideal [8]. Although the introduction of more image sensors (depth and thermal infrared) [9] can effectively reduce the influence of various interference factors, it also creates a lot of redundant information. The contradiction between accuracy and speed is also one of the main problems faced by the defect recognition algorithm [10]; in the actual production environment, the speed of the assembly line is relatively fast, for flat steel products, the assembly line running speed is about 20 m/s, for long wire, its product production assembly line running speed is up to 100 m/s. To achieve real-time detection, the algorithm developed must have a high enough accuracy at the same time, and the recognition speed is also fast enough. However, these two points are often contradictory in the design process of the algorithm. In the recognition of some small defects, it can be found that the defect area is too small in relation to the whole detection object, which leads to the detection object defect characteristics not being obvious. In addition, the manufacturing industry may be different for different steel mills, which may lead to different types of defects in their products, so the generalization of the algorithm will not be ideal, and it will be more difficult to produce a high-quality general dataset, which needs to go to different steel mills to collect samples. Finally, the collection of image samples of the inspected objects is also difficult, because for some steels, multiple planes need to be inspected, which leads to the need for multiple cameras to ensure complete coverage of the steel surface, so the collection of images from multiple cameras is also difficult.

The most common steel products include hot/cold rolled strip, bar/wire rod, slab, billet, and plate, which cover most of the uses of steel as a basic application material. The study of hot/cold rolled steel strips has received the most attention, mainly because most of these products are finished products. In terms of shape, the steel surface can be divided into flat products and long products [11]. Among them, flat products can be subdivided into slab/billet, steel plate, hot/cold rolled strip, and coated strip. Strip products include rods/wires, which are made by a hot rolling process with an oxidized surface, and other products such as rails, angle steel, channel steel, etc., which have a complex cross section, so the detection is also more complex. The specific division structure of the steel surface types is shown in Figure 1.

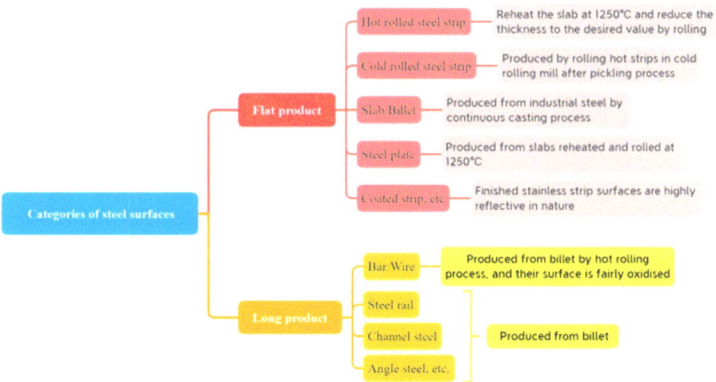

Figure 1. Categories of steel products.

The existence of different defects on the surface of different types of steel and the absence of an accepted standard for the classification of defects leads to the possibility of a certain irrationality in the classification of defects, which adds to the difficulties in their recognition. In this paper, the defect categories mainly detected in a large amount of literature are summarized, and the defect catalog published by Verlag Stahleisen GmbH [12] was used as a basis to summarize the defect categories contained in the steel surface for the types of the steel surface, and the specific defect categories are shown in Table 1.

Table 1. Statistics of the steel surface defect categories.

Steel Surface Type	Defect Category
Slab	Crack, pitting, scratches, scarfing defect
Plate	Crack, scratch, seam
Billet	Corner crack, line defect, scratch
Hot rolled steel strip	Hole, scratch, rolled in scale, crack, pits/scab, edge defect/coil break, shell, lamination, sliver
Cold rolled steel strip	Lamination, roll mark, hole, oil spot, fold, dark, heat buckle, inclusion, rust, sliver, scale, scratch, edge etc.
Stainless steel	Hole, scale, scratch, inclusion, roll mark, shell, blowhole
Wire/Bar	Spot, dark line, seam, crack, lap, overfill, scratch etc.

The steel surface defect recognition methods go through three stages: manual recognition, the traditional machine learning algorithm, and deep learning algorithm. The milestones regarding the development of industrial surface defect technology are shown in Figure 2.

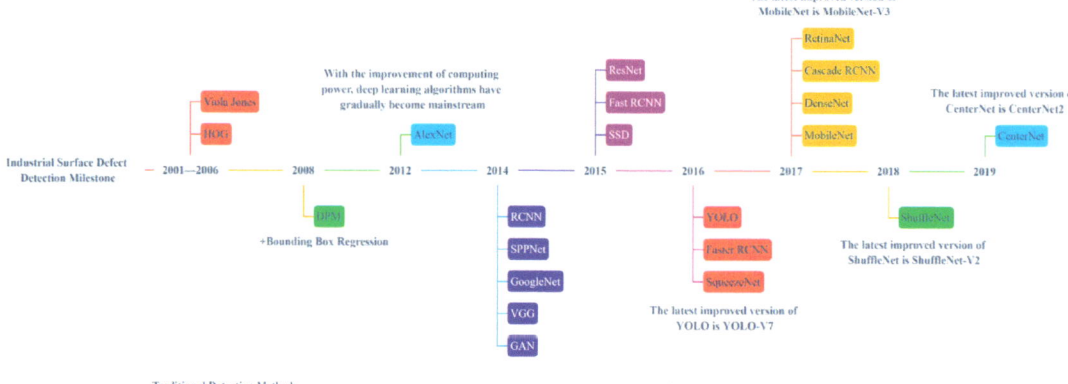

Figure 2. Development process of technologies related to industrial surface defect detection. Viola Jones [13], HOG [14], DPM [15], AlexNet [16], GAN [17], VGG [18], Google [19], SPPNet [20], RCNN [21], ResNet [22], Fast RCNN [23], SSD [24], SqueezeNet [25], Faster RCNN [26], YOLO [27], RetinaNet [28], Cascade RCNN [29], DenseNet [30], MobileNet [31], ShuffleNet [32], CenterNet [33].

The first stage of manual recognition methods refer to the detection of steel surface defects by quality inspectors, mainly by virtue of manual experience and subjective judgment: there are low efficiency, labor intensity, low precision, poor real-time, and other defects [34]. In addition, the human eye observation range and speed are limited, generally requiring that the product width cannot exceed 2 m and the product movement speed cannot exceed 30 m/s, otherwise the human eye will not be able to make effective observations [35], so manual defect recognition can no longer meet the increasingly developed industrial practical production environment. The second is the traditional machine learning algorithm, which generally needs to manually design the feature extraction rules of the recognition object, which is extremely dependent on the expert's knowledge judgment, so human experience plays a decisive role in these methods, resulting in poor robustness of the algorithm and insufficient generalization ability. Traditional machine learning methods generally combine feature extraction methods with classifiers such as SVM to realize defect recognition. Such methods are usually sensitive to defect size and noise, and are generally not end-to-end models that cannot meet the needs of automatic defect recognition [36]. In addition, conventional machine learning methods rely on certain assumptions to be valid [37], for example, the color depth of the defective part is not the same as the background part, which limits the scope of application of this type of algorithm. For deep learning algorithms, with the advancement in computing power, they have gradually started to become popular in the field of surface defect detection in recent years, and deep learning methods have gradually been introduced into steel surface defect recognition. These methods can autonomously extract image features in the dataset through neural network models, and the characterization capability is more powerful, avoiding the manual design of feature extraction rules, which can effectively overcome the shortcomings of traditional machine learning methods. However, deep learning methods also have some challenges in the actual application environment, as described in Section 5.

The rest of this paper is organized as follows. In Section 2, the key elements of the hardware architecture of the steel surface defect detection system will be discussed. In Section 3, the steel surface defect recognition algorithms are classified and summarized. In Section 4, the datasets and algorithm performance evaluation metrics used in recent papers are summarized. In Section 5, the existing challenges of the steel surface defect recognition algorithm are described, and suggestions for their solution are provided. In Section 6, the entire paper is summarized, and the future directions of our work are described.

2. Key Hardware for Steel Surface Defect Recognition System

A complete steel surface defect recognition system consists of three parts: image acquisition, defect recognition, and quality control. The image acquisition aims to measure and capture the image information of the object to be inspected using an optical system that consists of a camera or analog camera and an illumination system. The camera is used to capture the image of the object to be detected, while the light source is used to assist the camera in capturing a higher quality image, which is beneficial for improving the detection accuracy and training efficiency. The defect recognition process is implemented by the software algorithm, and the specific description of the defect recognition algorithm is described in Section 3. The defect recognition section delivers the recognition results to the quality control section, thus displaying the strengths and weaknesses of the detected object as well as key information such as the type and location of the defect, etc. The quality control section only requires the reading and display of the information, so no detailed description is needed. A diagram of the steel surface defect recognition system is shown in Figure 3.

2.1. Camera

Industrial cameras can be classified according to the output mode and can be divided into two categories: analog cameras and digital cameras. As the output signal form of analog camera is a standard analog video signal, it needs to be equipped with a special image acquisition card to be converted into digital information that can be processed by a computer. Analog cameras are generally used in the field of TV cameras and surveillance, with the characteristics of good versatility and low cost, but generally lower resolution, slow acquisition speed, and the image transmission is susceptible to noise interference, resulting in image quality degradation, so it can only be used for machine vision systems with low requirements for image quality. In contrast, digital cameras are integrated with internal A/D conversion circuit, which can directly convert the analog image signal into digital information, which not only effectively avoids the problem of interference in the image transmission line, but also generates higher quality images. Compared with analog cameras, digital cameras also have higher resolution and frame rate, smaller size, and less power consumption requirements [38]. Therefore, digital cameras are more suitable for working in fast steel production lines and complex environments in real steel mills than analog cameras.

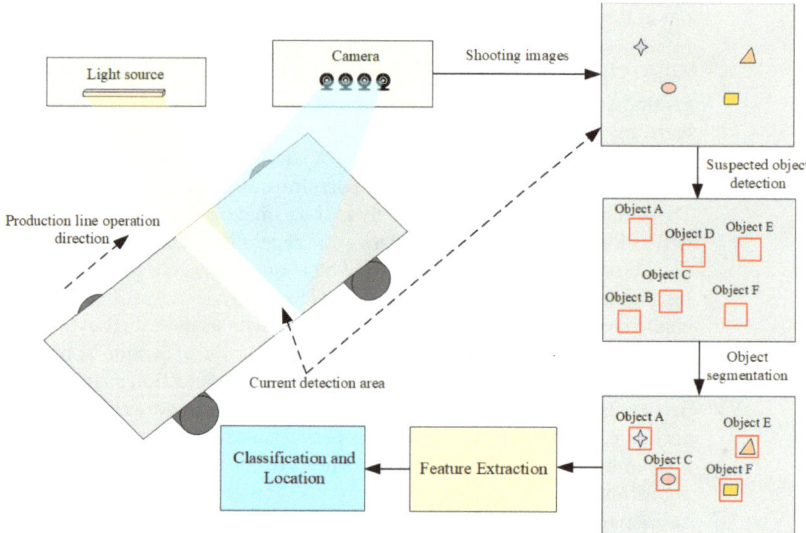

Figure 3. Steel surface defect recognition system.

If divided by chip type, industrial cameras can be further divided into two categories: charge coupled device (CCD) cameras and complementary metal oxide semiconductor (CMOS) cameras. The difference between these two cameras lies in the way that light is converted into electrical signals. For CCD sensors, light striking the image element generates an electric charge that is transmitted and converted into the current, buffer, and signal output through a small number of output electrodes. For CMOS sensors, each image element completes its own charge-to-voltage conversion while generating digital signals. CCD cameras do not have an absolute advantage over CMOS cameras because CCD sensors can adapt to the brightness range of 0.1–3 lux, which is 3 to 10 times higher than the general CMOS cameras, so the current general CCD cameras have a higher imaging quality than CMOS cameras. However, because the CMOS sensor has a photosensitive element, amplifier, A/D converter, memory, digital signal processor, and computer interface control circuit integrated in a silicon chip, it has a simple structure, is fast, has a low power consumption, low cost, and other characteristics. With the development of technology, the CMOS camera's poor image quality, small image sensitive unit size, and other problems have gradually been solved by the emergence of an "active image sensitive unit", which increased the ability to resist noise. This means that CMOS sensors have an almost comparable sensitivity with CCD sensors, the image quality can be improved, and in terms of power consumption, the processing speed is better than that of CCD sensors. Therefore, many people believe that CMOS will become the leading sensor technology for machine vision in the future [39]. In light of the actual production environment in steel mills, CMOS cameras are more suitable for future defect detection systems. CMOS sensors do not require complex processing and directly convert the electrons generated by the image semiconductor into a voltage signal, so they are very fast, an advantage that makes CMOS sensors very useful for high frame cameras, where high frame speeds can easily reach over a thousand frames at speed, which is very suitable for high-speed production lines.

In summary, since the steel surface defect recognition process and the capture of dataset images are generally performed on steel mill production lines, and the actual steel mill needs to face the complex and harsh inspection environment of vibration, light, high temperature, fast, steam and oil, etc., CMOS type digital cameras are recommended as the actual image capture tool.

2.2. Light Source

As the camera exposure time is relatively short when shooting at high speed, in order to let enough light into the camera in a short time, proper fill light is essential. A suitable lighting system can help the camera capture sharper images, thus making the entire inspection system more efficient and accurate. Light sources often used in machine vision are fluorescent, incandescent, xenon, and light emitting diode (LED). Among them, the LED light source is the most widely used in the field of steel defect recognition [40–42]. This is because the LED lighting cycle is long, can generally be illuminated for up to 100,000 h, and with its lower heat, low power consumption, uniform and stable brightness, and variety of colors can be made into a variety of shapes and sizes and can set a variety of irradiation angles to meet a variety of lighting needs. In addition, the LED light source response is fast, can reach maximum brightness in 10 microseconds or less, has a power supply with an external trigger, can be controlled by computer, has fast start, low operating costs, and long life LED, which will reflect greater advantages in terms of the comprehensive cost and performance. Guidelines for setting up the LED light source can be found in [43]. Since some steel surfaces are relatively smooth, specular reflections need to be avoided as much as possible, so diffusers can be added to the light set to reduce the glare reflected from the metal samples.

3. Algorithm Classification and Overview

Steel surface defect detection algorithms are classified according to whether deep learning techniques are applied and can be divided into traditional machine learning-based algorithms and deep learning-based algorithms. The traditional machine learning algorithms can be broadly classified into three categories: texture feature-based methods, color feature-based methods, and shape feature-based methods. The algorithms based on deep learning can be roughly divided into supervised methods, unsupervised methods, and weakly supervised methods. Of course, deep learning methods can also be classified according to the function of the selected neural network, which can be divided into image classification networks and object detection networks. Among them, image classification networks such as the classic AlexNet, ResNet, Visual geometry group (VGG), etc., and object detection networks can be divided into single-stage methods and two-stage methods, single-stage methods such as the well-known You only look once (YOLO) and Single shot multibox detector (SSD), and two-stage networks such as Fast R-CNN and Faster R-CNN.

3.1. Defect Recognition Algorithm Based on Traditional Machine Learning

The traditional machine learning approach was an epoch-making advancement from manual inspection, and usually starts with the manual design of feature extraction rules, followed by feature extraction, and finally feeds the extracted features into the classifier to achieve the classification of defects. Because of the reliance on manually designed feature extraction rules, it leads to poor robustness and generalization ability of the algorithm and is susceptible to interference and the influence of noise, thus reducing the detection accuracy. The most traditional methods basically only provide a defect classification function and do not perform defect localization or segmentation, which is an incomplete defect recognition process. The machine learning algorithms used for steel surface defect recognition can be broadly classified into texture feature-based methods, shape feature-based methods, and color feature-based methods. However, in the field of steel surface defect detection, since color features mainly refer to grayscale features of the image, and the methods used to extract grayscale features are statistically based, the color feature-based methods were classified here under the texture feature-based methods. An illustration of the classification based on traditional machine learning methods in the field of steel surface defect recognition is shown in Figure 4.

Figure 4. Classification of traditional machine learning methods.

3.1.1. Texture Feature-Based Methods

Texture feature-based methods are the most common methods in the field of steel defect detection, which reflects the homogeneity phenomenon in the image and can reflect the organization and arrangement characteristics of the image surface through the grayscale distribution of pixels and their nearby spatial neighborhood [44]. As shown in Figure 4, it can be subdivided into statistical-based methods, filter-based methods, structure-based methods, and model-based methods. These four methods can be used in combination or in conjunction with each other to achieve a higher performance. Regarding the literature on texture-based feature methods, these are shown in Table 2.

Statistical-based methods are used to measure the spatial distribution of pixel values, usually by using the grayscale distribution of image regions to describe texture features such as heterogeneity and directionality. Its common statistical methods include histogram, co-occurrence matrix, local binary patterns, etc. In 2015, Chu et al. [47] proposed a feature extraction method based on smoothed local binary patterns, which is insensitive to noise and invariant to scale, rotation, translation, and illumination, so the algorithm can maintain a high classification accuracy for the identification of strip surface defects. In 2017, Truong and Kim [48] proposed an automatic thresholding technique, which is an improved version of the Otsu method with an entropy weighting scheme that is able to detect very small defect areas. Luo et al. [49] proposed a selective local binary pattern descriptor, which was used to extract defect features, and then combined it with the nearest neighbor classifier (NNC) to classify strip surface defects; this algorithm pursued the comprehensive performance of recognition accuracy and recognition efficiency. The following year, Luo et al. [52] also proposed an improved generalized complete local binary pattern descriptor and two improved versions of the improved complete local binary pattern descriptor (ICLBP) and improved the complete noise-invariant local structure pattern (ICNLP) to obtain the surface defect features of the hot rolled steel strip, and then used the nearest neighbor classifier to achieve defect recognition classification, thus achieving high recognition accuracy. Zhao et al., in 2018 [51], designed a discriminative manifold regularized local descriptor algorithm to obtain steel surface defect features and complete matching by the manifold distance defined in the subspace to achieve the classification of defects in images. In 2019, Liu et al. [53] proposed an improved multi-block local binary pattern algorithm to extract the defect features and generate grayscale histogram vectors for steel plate surface defect recognition, and this work was able to recognize images at 63 FPS with a high detection accuracy at the same time.

Table 2. Summary of algorithms based on image texture features.

Category	Year	Ref.	Object	Function	Methods	Performance
Statistical Based Methods	2013	[45]	Hot rolled steel strip	Defect Classification	Local binary pattern	SNR = 40, ACC = 0.9893
	2014	[46]	Steel strip	Defect Classification	Co-occurrence matrix	ACC = 0.9600
	2015	[47]	Steel strip	Defect Classification	Local binary pattern	ACC = 0.9005
	2017	[48]	Steel strip	Defect Location	Auto threshold	-
	2017	[49]	Hot rolled steel strip	Defect Classification	Local binary pattern	ACC = 0.9762, FPS = 10
	2017	[50]	Steel	Defect Classification	Histogram, co-occurrence matrix	ACC = 0.9091
	2018	[51]	Steel	Defect Classification	Local descriptors	ACC = 0.9982, FPS = 38.4
	2018	[52]	Hot rolled steel strip	Defect Classification	Local binary pattern	TPR = 0.9856, FPR = 0.2900, FPS = 11.08
	2019	[53]	Plate steel	Defect Classification	Local binary mode, gray histogram	ACC = 0.9440, FPS = 15.87
Filter Based Methods	2012	[54]	Hot rolled steel strip	Defect Classification	Curved wave transform	ACC = 0.9733
	2015	[55]	Thick steel plate	Defect Detection	Gabor	ACC = 0.9670, FPR = 0.75
	2015	[56]	Continuous casting slabs	Defect Classification	Shearlet	ACC = 0.9420
	2017	[57]	Steel slabs	Defect Detection	Gabor	ACC = 0.9841
	2018	[58]	Hot rolled steel strip	Defect Classification	Shearlet	ACC = 0.9600
	2013	[59]	Hot rolled steel strip	Defect Detection	Wavelet transform	G-mean = 0.9380, Fm = 0.9040
	2014	[60]	Plate steel	Defect Detection	Gabor	TPR = 0.9446, FNR = 0.29
	2019	[61]	Continuous casting slabs	Defect Classification	Contour wave	AP = 0.9787
Structure Based Methods	2016	[41]	Steel rails	Defect Location	Edge	-
	2010	[62]	Steel strip	Defect Location	Edge	-
	2015	[63]	Steel strip	Defect Detection	Morphological operations	ME = 0.0818, EMM = 0.3100, RAE = 0.0834
	2016	[64]	Steel rails	Defect Location	Skeleton	ACC = 0.9473, FPS = 1.64
	2014	[65]	Silicon Steel	Defect Segmentation	Morphological operations	-
Model Based Methods	2018	[66]	Steel strip	Defect Segmentation	Guidance template	PRE = 0.9520, RECALL = 0.9730, Fm = 0.9620, FPS = 28.57
	2018	[67]	Steel sheet	Defect Segmentation	Low-rank matrix model	AUC = 0.835, Fm = 0.6060, MAE = 0.1580, FPS = 5.848
	2019	[68]	High strength steel joints	Defect Classification	Fractal model	ACC = 0.8833
	2013	[69]	Steel strip	Defect Segmentation	Markov model	CSR = 0.9440, WSR = 0.1880
	2019	[70]	Hot rolled steel	Defect Segmentation	Compact model	FPR = 0.088, FNR = 0.2660, MAE = 0.1430

* SNR indicates Gaussian noise, ME indicates misclassification error, EMM indicates edge mismatch, RAE indicates relative foreground area error, CSR indicates correct segmentation rate, WSR indicates wrong segmentation rate, FNR indicates false negative rate, FPR indicates false positive rate, MAE indicates mean square error, AUC indicates area under the curve, TPR indicates true positive rate, FPS indicates frames per second.

Filter-based methods are also called spectrum based methods and can be divided into spatial domain based methods, frequency domain methods, and space–frequency domain methods. They aim to treat the image as a two-dimensional signal, and then analyze the image from the point of view of signal filter design. The filter-based methods include curvelet transform, Gabor filter, wavelet transform, and so on. Xu et al. [54] achieved the multiscale feature extraction of surface defects of a hot-rolled steel strip by curvilinear wave transform and kernel locality preserving projections (KLPP), thus generating high-dimensional feature vectors before dimensionality reduction, and finally, defect classification by SVM. In 2015, Xu et al. [55] designed a scheme that introduced Shearlet transform to provide effective multi-scale directional representation, where the metal surface image is decomposed into multiple directional sub bands by Shearlet transform, thus synthesizing high-dimensional feature vectors, which were used for classification after dimensionality reduction. Doo-chul CHOI et al. [57] used a Gabor filter combination to extract the candidate defects and preprocessed them with the double threshold method to detect whether there were pinhole defects on the steel plate surface. In 2018 [58], the classification of surface defects of a hot-rolled steel strip was achieved by extracting multidirectional shear wave features from the images and performing gray-level co-occurrence matrix (GLCM) calculations on the obtained features to obtain a high-dimensional feature set, before finally using principal component analysis (PCA) for dimensionality reduction followed by SVM for defect classification. Liu et al. [61] improved the contour wave transform based on the contour wave transform and the non-downsampled contour wave transform, and combined the multi-scale subspace of kernel spectral regression for feature extraction to achieve a relatively good recognition speed and the algorithm is applicable to a wide range of metallic materials.

The core goal of structure-based methods is to extract texture primitives, followed by the generalization of spatial placement rules or modeling, which is based on texture primitive theory. Texture primitive theory indicates that texture is composed of some minimal patterns (texture primitive) that appear repeatedly in space according to a certain rule. This method is applicable to textures with obvious structural properties such as texture primitives such as density, directionality, and scale size. In 2014, Song et al. [65] used saliency linear scanning to obtain oiled regions and then used morphological edge processing to remove oil interference edges as well as reflective pseudo-defect edges to enable the recognition of various defects in silicon steel. In 2016, Shi et al. [41] reduced the effect of interference noise on defect edge detection by improving the edge detection Sobel algorithm, thus achieving accurate and efficient localization of rail surface defects. Liu et al. [63] proposed an enhancement operator based on mathematical morphology (EOBMM), which effectively alleviated the influence of uneven illumination and enhanced the details of strip defect images. In 2016, [64] applied morphological operations to extract features of railway images and used Hough transform and image processing techniques to detect the track images obtained from the real-time camera to accurately recognize defect areas and achieve real-time recognition.

Model-based methods construct a representation of an image by modeling multiple attributes of a defect [71]. Some of the more common model-based approaches in the field of industrial product surface defect recognition are Markov models, fractal models, Gaussian mixture models, and low-rank matrix models, etc. In 2013, Xv et al. [69] introduced an environment-based multi-scale fusion method CAHMT based on the hidden Markov tree model HMT to achieve multi-scale segmentation of strip surface defects, which greatly reduced the error rate of fine-scale segmentation and the complexity of the algorithm. In 2018 [67], a saliency detection model of double low-rank sparse decomposition (DLRSD) was proposed to obtain the defect foreground image. Finally, the Otsu method was used to segment the steel plate surface defects, which improved the robustness to noise and uneven illumination. In 2019, [66] detected strip surface defects based on a simple guidance template. By sorting the gray level of the image, the sorted test image was subtracted from the guidance template to realize the segmentation of the strip surface defects. In the same year, Wang et al. [70] constructed a compact model by mining the inherent prior of the image, which provided good generalization for different inspection tasks (e.g., hot-rolled strip, rails) and had good robustness.

A summary of the characteristics of the commonly used texture feature-based methods is shown in Table 3 for the reference of future researchers.

Table 3. Summary of methods based on the image texture features.

Category	Methods	Ref.	Advantages	Disadvantages
Statistical Based Methods	Threshold technology	[48]	Simple, easy to understand and implement.	It is difficult to detect defects that do not differ much from the background.
	Clustering	[49]	Strong anti-noise ability and high computational efficiency	Vulnerable to pseudo defect interference.
	Grayscale feature statistics	[50]	Suitable for processing low resolution images.	Low timeliness, no automatic threshold selection.
	Co-occurrence matrix	[46]	The extracted image pixel space relationship is complete and accurate.	The computational complexity and memory requirements are relatively high.
	Local binary pattern	[47]	Discriminative features with rotation and gray scale invariance can be extracted quickly.	Weak noise immunity, pseudo-defect interference.
	Histogram	[53]	Suitable for processing images with a large grayscale gap between the defect and the background.	Low detection efficiency for complex backgrounds, or images with defects similar to the background.
Filter Based Methods	Gabor filter	[55]	Suitable for high-dimensional feature spaces with low computational burden.	Difficult to determine optimal filter parameters and no rotational invariance.
	Wavelet filters	[59]	Suitable for multi-scale image analysis, which can effectively compress images with less information loss.	Vulnerable to correlation of features between scales.
	Multi-scale geometric analysis	[56]	Optimal sparse representation for high-dimensional data, capable of handling images with strong noise background.	The problem of feature redundancy exists.
	Curvelet transform	[54]	High anisotropy with good ability to express information along the edges of the graph.	Complex to implement and less efficient.
	Shearlet and its variants	[58]	Multi-scale decomposition and the ability to efficiently capture anisotropic features.	Difficult to retain original image detail information.
Structure Based Methods	Edge	[41]	It is suitable for extracting some low-order features of the image and is easy to implement.	Vulnerable to noise and only suitable for low resolution images.
	Skeleton	[64]	Almost distortion less representation of the geometric and topological properties of objects.	Unsatisfactory image processing for complex backgrounds.
	Morphological operations	[63]	Great for random or natural textures, easy to calculate.	Only for non-periodic image defects.
Model Based Methods	Gaussian mixture model	[66]	Correlation between features can be captured automatically.	Large computational volume and slow convergence, sensitive to outliers.
	Fractal model	[68]	The overall information of an image can be represented by partial features.	Unsatisfactory detection accuracy and limitation for images without self-similarity.
	Low-rank matrix model	[67]	Strong discriminatory ability and adaptive nearest neighbor.	Unsatisfactory detection accuracy.
	MRF model	[69]	Can combine statistical and spectral methods for segmentation applications to capture local texture orientation information.	Cannot detect small defects. Not applicable to global texture analysis.

3.1.2. Shape Feature-Based Methods

Shape feature-based methods are also very effective defect detection methods. These methods obtain image features through shape descriptors, so the accuracy of the shape description becomes the key to the merit of the image defect recognition algorithm. A good shape descriptor should have the characteristics of geometric invariance, flexibility, abstraction, uniqueness, and completeness. The commonly used shape descriptors can be divided into two categories: one is the contour shape descriptor, which is used to describe the outer edge of the object area, and the other is the area shape descriptor, which is used to describe the whole object area. The common methods based on contour shape descriptors are Fourier transform and Hough transform, etc. For the method using Fourier transform, it mainly uses the closure and periodicity of the region boundary to convert the two-dimensional problem into a one-dimensional problem. For example, Yong-hao et al [72] enables the detection of longitudinal cracks on the surface of the continuous casting plate in a complex background by calculating the Fourier magnitude spectrum of each sub-band to obtain features with translational invariance. In addition, Hwang et al. [73] used linear discriminant analysis using short-time Fourier transform pixel information generated from ultrasound guided wave data to achieve defect detection on 304SS steel plates. The Hough transform methods use the global features of the image to connect the edge pixels to form a regionally closed boundary. For example, Wang et al. in 2019 [74] achieved the detection of product surface defects by using the fast Hough transform in the region of interest (ROI) extraction stage to detect the boundary line of the light source. Regional shape features include the length and width, elongation, area ratio, and other aggregate shape parameter methods, which is a simple shape expression method. In addition, moments are a more reliable and complex region shape feature including geometric moments, central moments, etc. As Hu invariant moments [75], moment expressions are commonly used to describe the shape of steel surface defect regions. As Hu et al. [76] used both Fourier descriptors and moment descriptors to extract the shape features of steel strip surface defect images, in addition to the grayscale features and geometric features of the images, and finally support vector machine (SVM) was used to classify the defects in the steel strip surface images. For shape feature extraction, it must be built on image segmentation and is extremely dependent on the accuracy of image segmentation. For both methods, based on texture features and shape features, they can also be used in combination. For example, Hu et al. [77] proposed a classification model based on the hybrid chromosome genetic algorithm (HCGA) and combined geometric, shape, texture and grayscale features to identify and classify steel strip surface defects.

3.2. Defect Recognition Algorithm Based on Deep Learning

With the increase in computing power and the excellent performance of deep learning methods, many researchers have applied deep learning methods to various industrial inspection scenarios, and they have become the mainstream defect detection methods. Compared with traditional machine learning methods, deep learning methods can extract deeper and more abstract image features through operations such as convolution and pooling, and thus have more powerful characterization capabilities and do not require human-designed feature extraction rules, allowing for end-to-end model design. Convolutional neural networks use convolutional operations to extract features from input images, which can capture different levels of semantic information, thus effectively learning feature representations from a large number of samples and making the model have more powerful generalization capabilities. In addition, CNNs using pooling layers and sparse connections can reduce the model parameters while ensuring the efficiency of computational resources and network performance [78]. A detailed classification based on deep learning methods is shown in Figure 5.

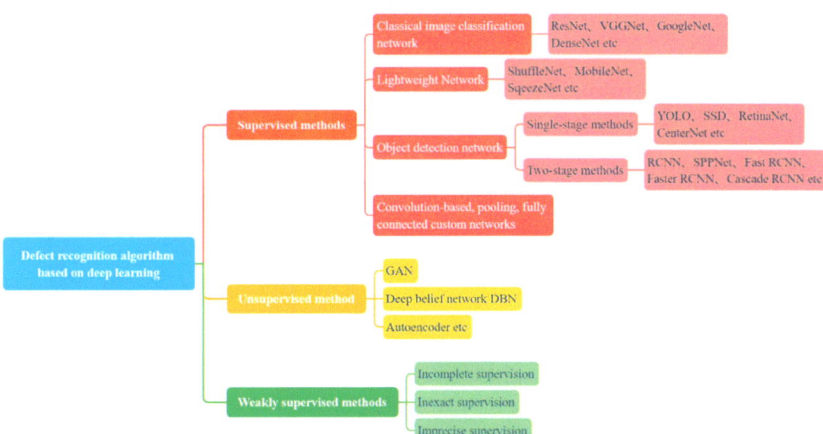

Figure 5. Defect recognition algorithm based on deep learning.

3.2.1. Supervised Methods

Supervised methods of deep learning require well-labeled training sets and test sets to verify the performance of the methods, which are generally more stable and accurate than other types of methods. However, the quality of the training set directly affects the performance of the algorithm and in a real industrial defect detection scenario, it is very difficult to produce a well-labeled and large dataset. Fu et al. [79] proposed a fast and robust lightweight network model based on SqueezeNet, which emphasizes the learning of low-level features and adds an MRF module, thus achieving accurate recognition of defect types using a small number of defect samples, and it is worth mentioning that the recognition efficiency of this work exceeded 100 fps. The classification-first framework proposed by He et al. [80] in 2019 consists of two networks: the classification network MG-CNN and YOLO, where the MG-CNN is used to detect defect categories, and then the set of feature maps with defects present is fed to the YOLO network to determine the defect locations based on the results of the classification. In 2020, Ihor et al. [81] verified through experiments that the pre-trained model ResNet50 was the best choice as a classification network for detecting steel surface defects and used binary focus loss to alleviate the problem of data sample imbalance to realize the recognition of steel surface defects. Li et al. [82] designed a scheme combining domain adaptive and adaptive convolutional neural networks, DA-ACNN, for the identification of steel surface defects. In 2021, Feng et al. [83] adopted a scheme combining the RepVGG algorithm and spatial attention mechanism to realize the recognition of surface defects of a hot rolled strip. A new defect dataset X-SDD for a hot rolled strip was proposed. However, the recognition efficiency was not high because of the large number of parameters. In the same year, [4] trained Unet and Xception separately as classifiers to detect surface defects on rolled parts using synthetic datasets, and the normal dataset training was used as a reference to verify the feasibility and effectiveness of manually generated datasets. In 2021, Wang et al. [35] improved the VGG19 model, which is shown in Figure 6, and the scheme was divided into two parts: online detection and offline training. The online part extracts the ROI regions of the defect images using the improved grayscale projection algorithm, and then detects the strip surface defects using the improved VGG19 model; the offline part adds the extracted ROI regions to the defect dataset and performs ROI image augmentation, adds the results to the balanced mixed dataset, and then uses the mixed dataset training to improve the performance of VGG19, thus effectively solving the problem of few samples or an unbalanced dataset.

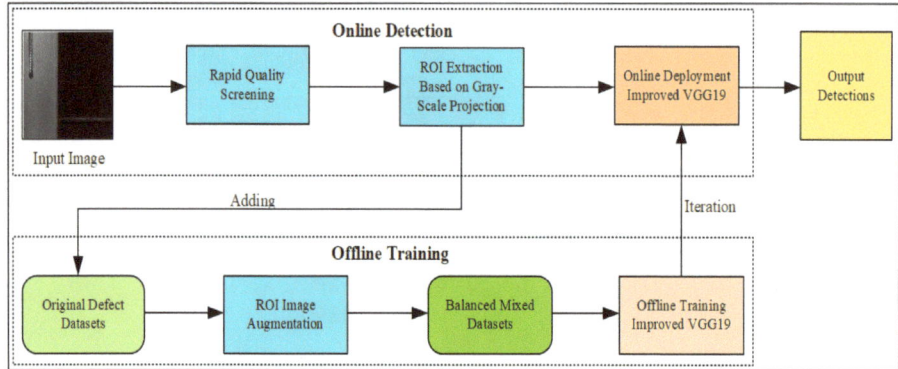

Figure 6. Model outline diagram [35].

Pan, Y et al. [37] incorporated the dual-attention module into DeepLabv3+, and Xception was selected as the network to achieve accurate to pixel-level defect localization segmentation. The edge-aware multilevel interaction network proposed by Zhou et al. in 2021 [84] used the ResNet network as the base backbone and adopted a U-shaped architecture composed of encoder decoders to recognize strip surface defects. In 2022, Liu et al. [85] designed the lightweight network CASI-Net, whose structure is shown in Figure 7. The network uses a lightweight feature extractor to extract image features, and then uses a collaborative attention mechanism and self-interaction module based on a biological vision to modify the features. Finally, multilayer perceptron (MLP) was used to classify the surface defects of the steel strip.

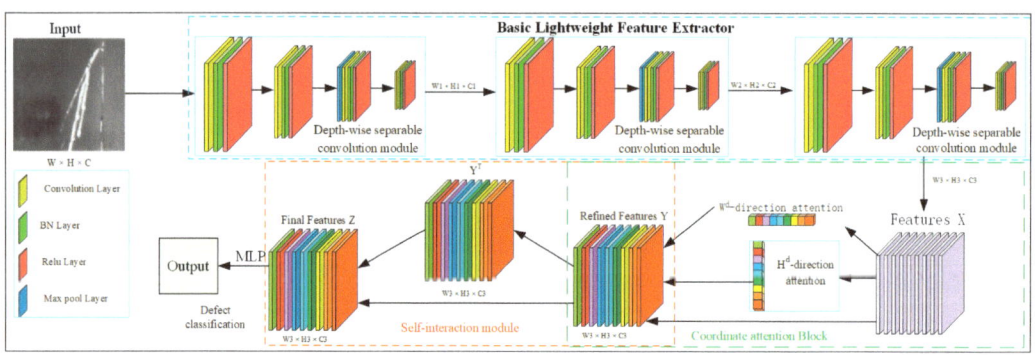

Figure 7. Overview of the CASI-Net framework [85].

3.2.2. Unsupervised Methods

Unsupervised methods, as opposed to supervised methods, do not require their training sets to be labeled, or are even just trained with enough defect-free samples, making up for the disadvantage that supervised methods have difficulty in producing datasets. However, in contrast, the detection accuracy of unsupervised methods is generally lower than that of supervised methods, and there is instability in the training results. The commonly used unsupervised methods include autoencoder, GAN, deep belief network [86,87], and self-organizing graph [88], etc. In 2017, Liu et al. [89] used an anisotropic diffusion model to eliminate the interference of pseudo-defects, then proposed a new HWV unsupervised model to characterize the texture distribution of each local block in the image, and finally invoked the adaptive thresholding technique to achieve the segmentation of the defects and background. Mei et al. (2018) [90] also used only defect-free samples to train the model, and the method was constructed based on the convolutional denoising autoencoder (CDAE) architecture of Gaussian pyramids to distinguish between defective and defect-free parts. In 2019, [91] proposed a single classification method based on GAN for steel strip surface defect detection, which could only detect the presence or absence of defects and cannot distinguish the categories. In the same year, [92] used convolutional autoencoder convolutional autoencoder (CAE) and sharpening processing to extract the defect features of the missing input image, and finally used Gaussian blurring and thresholding as post-processing to clarify the defects to achieve the segmentation of defects. In 2020, Niu et al. [93] proposed a global low-rank non-negative reconstruction algorithm with background constraints to fuse the detection results of 2D significant maps and 3D contour information to achieve the detection of rail surface defects.

3.2.3. Weakly Supervised Methods

Weakly supervised methods between supervised and unsupervised methods, which have three categories, namely, incomplete supervision, inexact supervision, and imprecise supervision [94]. The first two of these methods have both been well validated in the field of surface defect detection. Among them, inexact supervision refers to the use of a small amount of labeled data mixed with a large amount of unlabeled data for training, so that a more desirable accuracy can be obtained while avoiding the difficulty of producing a large number of labeled datasets. Uncertain supervision is the use of a small amount of fully labeled data (pixel-level labeling) mixed with a large amount of weakly labeled data (image-level or box-level), which showed a performance almost as good as that of fully supervised methods and reduced the trouble of producing fully labeled datasets. Finally, inaccurate supervision means that the given data samples may contain partially incorrect labeling information. The scheme proposed by He et al. [95] in 2019 combined a convolutional autoencoder with a semi-supervised GAN and introduced a passthrough layer in the CAE to extract fine-grained features, resulting in excellent recognition accuracy. In 2019, the Google Institute [96] designed a new algorithm MixMatch by unifying the current mainstream semi-supervised algorithms, and conducted extensive detection experiments using this algorithm, which will have a significant performance improvement compared with other weakly supervised methods. He et al. [97] proposed a multiple learning algorithm based on the GAN and ResNet18 networks, which could generate data samples by itself and provide labels for the samples, enabling the expansion of the dataset, thus further enhancing the recognition of defects with few samples. Jong et al., 2020 [98] proposed a new convolutional variational autoencoder, convolutional variational autoencoder (CVAE), which was used to generate defect images and then used these images to train the proposed CNN classifier to achieve high accuracy defect detection. The model designed by Jakob et al. (2021) [99] had two sub-network architectures [100,101], a segmentation sub-network that learns from pixel-level labels, and a classification sub-network that learns from weak image-level labels and combines these two networks to achieve hybrid supervision, and experimentally demonstrated that hybrid supervised training with only a few fully annotated samples added to weakly labeled image samples could yield comparable performance to the fully supervised model. Zhang et al. [102] proposed a weakly supervised learning method named CADN, which was implemented by extracting the category perception spatial information from the classification pipeline, only used weak image labels for training, and could simultaneously realize image classification and defect localization.

To facilitate the understanding of the characteristics of the three deep learning methods, a pipeline diagram of the deep learning methods is summarized in Figure 8 and the characteristics of the deep learning methods are shown in Table 4.

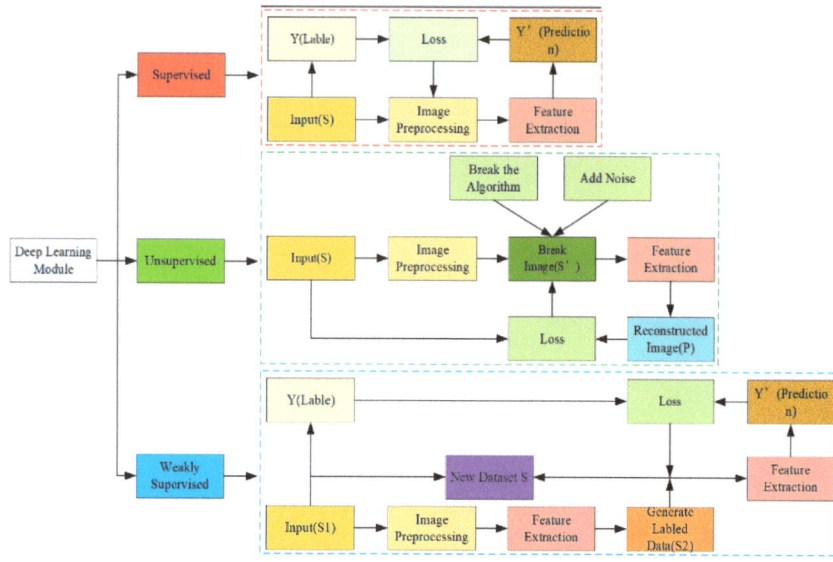

Figure 8. Diagram of the outline of the deep learning methods.

Table 4. Summary of the features of deep learning methods.

Category	Advantages	Disadvantages
Supervised methods	High precision, good adaptability, wide range of applications.	Dataset annotation is heavy and difficult to make.
Unsupervised methods	It can be trained directly using label-free data with simple techniques.	Relatively low precision, unstable training results are easily affected by noise and initial parameters.
Weakly supervised methods	It has the advantages of both supervised and unsupervised methods.	The training process is tedious and the technical implementation is complicated.

3.3. Object Detection Methods

The counterpart of the image classification network is the object detection network, which mainly contains single-stage methods and two-stage methods, each of which has its own advantages and disadvantages. The main difference between the single-stage and two-stage methods is whether there is a clear candidate box generation stage within the algorithm. The single-stage object detection algorithm directly calculates the image to generate the detection result, which is fast, but the detection accuracy is relatively low; the two-stage object detection algorithm first extracts the candidate box from the image, and then conducts the secondary correction based on the candidate region to obtain the detection point result, which has a higher detection accuracy, but slower detection speed. Therefore, what needs to be done in the field of object detection is to continuously optimize the mainstream object detection algorithms to achieve the best balance of detection accuracy and speed.

3.3.1. Single-Stage Methods

The most popular single-stage algorithms include SSD [24], YOLO [27], RetinaNet [28], CenterNet [33], etc. Li et al. [103] used a fully convolutional YOLO network to achieve the classification and localization of surface defects in steel strip with high recognition accuracy and speed. Yang et al. [104] proposed a deep learning algorithm for the defect detection of tiny parts based on a single short detector network SSD and speed model with a maximum detection accuracy of 99%. In 2021, Cheng et al. [105] proposed a steel surface defect recognition scheme based on RetinaNet with differential channel attention and adaptive spatial feature fusion (ASFF). Kou et al. [106] proposed an end-to-end defect detection model based on YOLO-V3 and combined it with anchor free feature selection to reduce the computational complexity of the model. However, this work could only detect defects in the normal size range, and very small defects in high-resolution images could not be detected. In 2022, Chen et al. [107] proposed a real-time surface defect detection method based on YOLO-V3 with a lightweight network MobileNeV2 selected for its backbone network, an extended feature pyramid network (EFPN) was proposed to detect multi-size objects, a feature fusion module was also designed to capture more regional details, and the scheme achieved high detection speed and accuracy. Tian et al. [108] proposed a steel surface defect detection algorithm called DCC-CenterNet, which not only focused on the center of the defect, but also extracted the overall information without drawing false attention.

3.3.2. Two-Stage Methods

In the field of industrial defect detection, common two-stage classical methods include RCNN [21], SPPNet [20], Fast RCNN [23], Faster RCNN [26], and Cascade RCNN [29], etc. The method proposed by Rubo et al. in 2020 [109] was based on Faster R-CNN and proposed three important improvements of weighted ROI pooling, FPN-based multi-scale feature extraction network, and strict-NMS to achieve more accurate defect recognition. Wang et al. in 2021 [36] proposed a combination of ResNet50 and an improved Faster R-CNN algorithm to detect steel surface defects and proposed three important improvements to the Faster R-CNN with spatial pyramid pooling (SPP), enhanced feature pyramid network (FPN), and matrix NMS algorithm to achieve higher detection accuracy. Zhao et al. [110] proposed an improved Faster R-CNN based network model, whose model structure is shown in Figure 9, where ResNet50 is used as the feature extraction network. First, the ResNet-50 network is reconstructed using deformable convolution. Second, the feature pyramid network is used to fuse multi-scale features and replace the fixed region of interest pooling layer with a variable pooling layer. Finally, the soft non-maximum value suppression algorithm (SNMS) is used to suppress detection frames that have significant overlap with the highest score detection frames, thereby enhancing the network's ability to identify defects. The method proposed by Li et al. [8] in 2022 is based on the improved YOLOV5 and Optimized-Inception-ResNetV2 models, where the first stage is to locate defects with the improved YOLOV5, and the second stage is to extract defect features and classify them with Optimized-Inception-ResNetV2. A summary of deep learning-based methods for identifying steel surface defects is shown in Table 5.

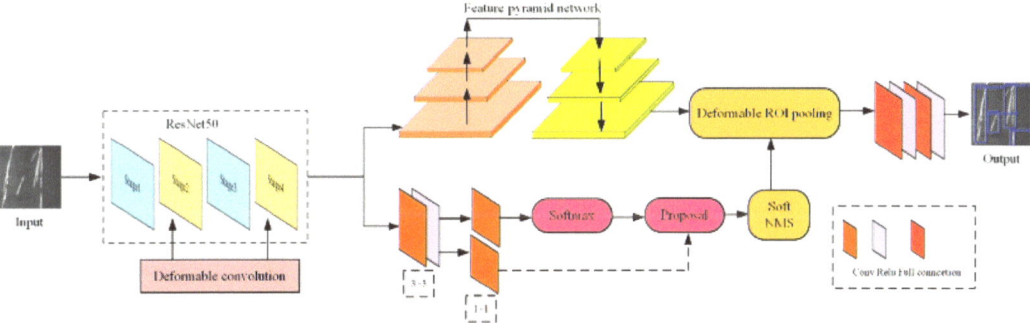

Figure 9. The improved method of Faster R-CNN [110].

Table 5. Summary of deep learning-based algorithms.

Category	Year	Ref.	Methods	Object	Function	Performance
Supervised Methods	2017	[111]	CNN	Metal	Defect Classification	ACC = 0.9207
	2017	[112]	Decay	Multi-Type	Defect Detection	ACC = 0.9400, FPS = 17, EE = 0.2100
	2019	[113]	VGG + LSTM	Steel plate	Defect Detection	ACC = 0.8620
	2019	[114]	Du-Net	Metal	Defect Segmentation	ACC = 0.8345
	2019	[115]	InceptionV4	Hot rolled Steel	Defect Classification	RR = 0.9710
	2019	[79]	SqueezeNet	Steel	Defect Classification	ACC = 0.9750, FPS = 100, Model size = 3.1 MB
	2019	[80]	MG-CNN	Hot rolled Steel	Defect classification and Location	CR = 0.9830, DR = 0.9600
	2020	[81]	ResNet50	Steel	Defect Classification	PRE = 0.8160, ACC = 0.9670, F1 = 0.6610, RECALL = 0.5670
	2021	[82]	DA-ACNN	Steel	Defect Classification	ACC = 0.9900
	2021	[83]	RepVGG	Hot rolled steel strip	Defect Classification	ACC = 0.9510, RCALL = 0.9392, PRE = 0.9516, F1 = 0.9325, Params = 83.825 M
	2021	[4]	Unet + Xception	Rolled piece	Defect Classification and Segmentation	PRE = 0.8400, RECALL = 0.9000, Dice score = 0.5950
	2021	[35]	VGG19	Steel strip	Defect Classification	ACC = 0.9762, FPS = 52.1
	2021	[37]	DAN-DeepLabv3+	Steel	Defect Precise Segmentation	mIoU = 0.8537, PRE = 0.9544, RECALL = 0.9071, F1 = 0.9297
	2021	[84]	ResNet34	Steel strip	Defect Precise Seg-mentation	MAE = 0.0125, WF = 0.9200, OR = 0.8380, SM = 0.9380, PFOM = 0.9120, FPS = 47.6
	2022	[85]	CASI-Net	Hot rolled steel strip	Defect Classification	ACC = 0.9583, Params = 2.22 M
Unsupervised Methods	2020	[93]	GLRNNR	Steel rails	Defect Detection and Segmentation	MAE = 0.0900, AUC = 0.9400, PRE = 0.9481, RECALL = 0.8066, Fm = 0.8716
	2017	[90]	MSCDAE	Multi-Type	Defect Detection and Segmentation	RECALL = 0.6440, PRE = 0.6400, FA = 0.6380
	2019	[92]	CAE	Hot rolled steel strip	Defect Segmentation	–
	2018	[116]	FCAE	Multi-Type	Defect Segmentation	PRE = 0.9200, FPS = 12.2
	2019	[91]	GAN	Steel strip	Defect Detection	PRE = 0.9410, RECALL = 0.9380, Fm = 0.9390
	2017	[89]	HWV	Steel	Defect Segmentation	FPS = 19.23, PRE = 0.9570, RECALL = 0.9680, Fm = 0.9620
Weakly Supervised Methods	2019	[40]	GAN	Multi-Type	Defect Classification and Segmentation	RECALL = 0.8710, ACC = 0.9920, AUC = 0.9140
	2019	[95]	CAE + GAN	Steel	Defect Classification	CR = 0.9650
	2019	[117]	D-VGG16	Multi-Type	Defect Classification and Segmentation	AP = 0.9913, PR = 0.9836, TPR = 0.9967, FPR = 0.0164, FNR = 0.0033
	2019	[97]	GAN + ResNet18	Steel	Defect Classification	ACC = 0.9507
	2020	[102]	CAND	Multi-Type	Defect Classification and Segmentation	ACC = 0.8910, PRE = 0.5510, RECALL = 0.9200, F1 = 0.6900, mAP = 0.6120
	2020	[98]	CVAE	Metal	Defect Classification	ACC = 0.9969, F1 = 0.9971
	2021	[99]	Dual network model	Steel	Defect Classification and Segmentation	AP = 0.9573
Single-stage Methods	2018	[103]	YOLO	Steel strip	Defect Classification and Location	ACC = 0.9755, FPS = 83, mAP = 0.9755, RECALL = 0.9586
	2020	[118]	YOLOV3-Dense	Steel strip	Defect Classification and Location	mAP = 0.8273, FPS = 103.3, F1 = 0.8390
	2021	[105]	RetinaNet	Steel	Defect Classification and Location	mAP = 0.7825, FPS = 12, FLOPs = 105.3, Params = 42.2
	2021	[106]	YOLOV3	Steel strip	Defect Classification and Location	mAP = 0.7220, FPS = 64.5
	2022	[107]	YOLOV3	Hot rolled steel strip	Defect Classification and Location	PRE = 0.9837, RECALL = 0.9548, F1 = 0.9690, mAP = 0.8696, FPS = 80.96
	2022	[108]	Center Net	Steel	Defect Classification and Location	mAP = 0.7941, FPS = 71.37

Table 5. *Cont.*

Category	Year	Ref.	Methods	Object	Function	Performance
Two-stage Methods	2020	[119]	SSD + Resnet	Steel	Defect Classification and Location	PRE = 0.9714, RECALL = 0.9214, Fm = 0.9449
	2020	[109]	Faster RCNN	Steel	Defect Classification and Location	DR = 0.9700, FDR = 0.1680
	2021	[36]	Faster RCNN	Steel	Defect Classification and Location	ACC = 0.9820, FPS = 15.9, F1 = 0.9752
	2021	[110]	Faster RCNN + FPN	Steel	Defect Classification and Location	mAP = 0.7520
	2022	[8]	YOLOV5 + Optimized-Inception-ResNetV2	Hot rolled steel strip	Defect Classification and Location	mAP = 0.8133, FPS = 24, Param = 37.7, RECALL = 0.7630

* RR indicates recognition rates, CR indicates classification rate, DR indicates detection rate, OR indicates overlapping ratio, FDR indicates false detection rate.

4. Datasets and Performance Evaluation Metrics

4.1. Datasets

Datasets are the basis of research work on steel surface defect recognition. A good dataset is more conducive to problem identification, thus facilitating problem solving. With the improvements at the industrial manufacturing level, the number of defective products is becoming smaller and smaller, so it is very challenging to create a high-quality dataset for the training of aa defect recognition algorithm. In addition, the labeling of data samples is also more labor-intensive. A high-quality dataset is very important for the defect recognition algorithm, which will directly affect the final performance of the algorithm. Therefore, the commonly used publicly available datasets have been summarized in the field of steel surface defect recognition. Most of the selected datasets were steel surface defect datasets, but they also contain some texture datasets of various materials for future researchers. A summary of the datasets in the field of industrial product defect detection is shown in Table 6.

Table 6. Summary of the datasets.

Dataset	Object	Description	Link
NEU [45]	Hot rolled steel strip	1800 grayscale images of hot-rolled strip containing six types of defects, 300 samples of each.	http://faculty.neu.edu.cn/songkc/en/zdylm/263265 (accessed on 9 November 2022)
Micro Surface Defect Database [120]	Hot rolled steel strip	Microminiature strip defect data, with defects only about 6 × 6 pixels in size.	http://faculty.neu.edu.cn/songkc/en/zdylm/263266 (accessed on 9 November 2022)
X-SSD [83]	Hot rolled steel strip	7 typical defects of hot-rolled steel strip, with 1360 defect images.	https://github.com/Fighter20092392/X-SDD-A-New-benchmark (accessed on 9 November 2022)
Oil Pollution Defect Database [65]	Silicon Steel	Oil-disturbed silicon steel surface defects dataset	http://faculty.neu.edu.cn/songkc/en/zdylm/263267 (accessed on 9 November 2022)
Severstal: Steel Defect Detection	Steel plate	There are 12,568 grayscale images of steel plates of size 1600 × 256 in the training dataset, and the images are divided into 4 categories.	https://www.kaggle.com/c/severstal-steel-defect-detection/data (accessed on 9 November 2022)
UCI Steel Plates Faults Data Set [121]	Steel strip	This dataset contains 7 types of strip defects. This dataset is not image data, but data of 28 features of strip defects.	https://archive-beta.ics.uci.edu/dataset/198/steel+plates+faults (accessed on 2 May 2022)

Table 6. Cont.

Dataset	Object	Description	Link
SD-saliency	Steel strip	Contains a total of 900 cropped images containing 3 types of defects, each with a resolution of 200 × 200.	https://github.com/SongGuorong/MCITF/tree/master/SD-saliency-900 (accessed on 9 November 2022)
GC10-DET [122]	Steel strip	The dataset contains 2257 images of steel strip with 10 defect types and an image resolution of 4096 × 1000	https://github.com/lvxiaoming2019/GC10-DET-Metallic-Surface-Defect-Datasets (accessed on 2 May 2022)
RSDDs Dataset [123]	Steel rails	Two types of orbital surface images (67 images and 128 images)	http://icn.bjtu.edu.cn/Visint/resources/RSDDs.aspx (accessed on 2 May 2022)
DAGM [124]	Multi-Type	Includes 10 different computer-generated grayscale images of surfaces containing various defects.	https://hci.iwr.uni-heidelberg.de/node/3616 (accessed on 2 May 2022)
KolektorSSD2 [99]	Multi-Type	This dataset training set test set contains a total of 3335 color images, more than 5 kinds of defects.	https://www.vicos.si/resources/kolektorsdd2/ (accessed on 2 May 2022)
Kylberg Texture Dataset [125]	Multi-Type	The dataset contains 28 texture classes, each with 160 unique texture patches.	http://www.cb.uu.se/~gustaf/texture/ (accessed on 2 May 2022)

4.2. Defect Recognition Algorithm Performance Evaluation Metrics

The performance evaluation metrics of defect recognition algorithms were used to measure the performance of the designed algorithms, and the selection of the algorithm performance evaluation metrics should be comprehensive because the selection of different evaluation metrics may present different results. This paper summarizes some of the algorithm performance evaluation metrics commonly used in the field of steel defect detection, which can be broadly divided into two categories: the precision class metrics and the efficiency class metrics.

4.2.1. The Precision Class Metrics

The precision class evaluation metrics were used to evaluate the precision of the relevant algorithms for the classification of defect categories and the precision of the defect localization segmentation.

The first and most basic evaluation metrics are TP, TN, FP, and FN, where TP indicates the true positive, the number of correctly classified positive samples; TN indicates the true negative, the number of correctly classified negative samples; FP indicates the false positive, the number of negative samples classified as positive; and FN indicates the false negative, the number of positive samples classified as negative. Based on the above four basic evaluation metrics, evaluation metrics such as PRE, RECALL, ACC, false escape rate, and false alarm rate evolved. Among them, PRE represents the ratio of the number of correctly predicted positive samples to the number of predicted positive samples, the classification precision of the algorithm, with the following equation:

$$PRE = \frac{TP}{TP + FP}, \quad (1)$$

A higher PRE means a smaller random error, a smaller variance, describing the perturbation of the prediction results. RECALL indicates the ratio of the number of correctly predicted positive samples to the number of actual positive samples, which is given by the following formula:

$$RECALL = \frac{TP}{TP + FN}, \quad (2)$$

A higher value of RECALL means that the algorithm is more capable of detecting the object algorithm. ACC indicates the proportion of the total number of correctly predicted samples to the total number of samples, which is given by the following formula:

$$ACC = \frac{TP + TN}{TP + TN + FP + FN}, \quad (3)$$

A higher value of ACC represents a smaller systematic error, a smaller deviation, and describes the degree of deviation of the predicted result from the actual value. For the error escape rate, EE is defined as the number of negative samples judged to be positive as a proportion of the total number of negative samples in the sample, calculated as follows:

$$EE = \frac{FP}{TN + FP}, \quad (4)$$

For the false alarm rate, FA is defined as the ratio of the number of positive samples incorrectly determined as negative to the number of all positive samples, calculated as follows:

$$FA = \frac{FN}{TP + FN}, \quad (5)$$

It should be noted that the above metrics, if the values of the evaluation metrics of a single category are calculated in the multi-classification algorithm, it is necessary to take any of the classes in the multi-classification as a positive sample and combine the other classes into one class as a negative sample, so that the TP, TN, FP, and FN of each class can be calculated separately, and thus the other evaluation metrics of each class can be calculated. The formula of PRE and RECALL shows that the higher the PRE indicator, the higher the accuracy of predicting positives, but this indicator does not take into account the wrong prediction of negative samples, while the RECALL indicator is exactly the opposite situation, that is, it does not take into account the wrong prediction of positive samples, so it is possible that there is an algorithm with very high and very low PRE and RECALL metrics. A good algorithm also needs to have both a high PRE and RECALL. Therefore, a metric that integrates PRE and RECALL is needed, and the F-measure can meet this requirement with the following formula:

$$F - measure = (1 + \beta^2)\frac{PRE * RECALL}{\beta^2 PRE + RECALL}, \quad (6)$$

When $\beta = 1$, the F-measure is the F1-score. The evaluation metric WF is the weighted F-measure, which is calculated as follows:

$$WF = (1 + \beta^2)\frac{PRE^w \times RECALL^w}{\beta^2 PRE^w + RECALL^w}, \quad (7)$$

For the AP and mAP metrics, they are also calculated based on PRE and RECALL, where the AP average precision is defined by calculating the area under the P–R curve, and by changing different confidence thresholds, multiple pairs of PRE and RECALL, values can be obtained where the P–R curve is defined by putting the RECALL value on the X-axis and the PRE value on the Y-axis, resulting in a PRE–RECALL curve, referred to as the P–R curve. For the AP calculation method, the average accuracy calculation method is generally calculated by 11-point interpolation, and the calculation formula is as follows:

$$AP = \frac{1}{11} \times \sum_{i \in \{0, 0.1, \ldots, 1\}} AP_r(i), \quad (8)$$

Once the AP for all categories have been calculated, the value of the average precision mAP for all categories can be calculated with the following formula:

$$mAP = \frac{\sum_{i=1}^{K} AP_i}{K}, \qquad (9)$$

where K refers to the number of categories. Similar to the AP calculation is the AUC, which means the area under the ROC curve, and a higher value of AUC means a better effect of the corresponding classifier. The ROC curve is also called the perceptivity curve, and the horizontal coordinate of the curve is the FPR, which is the false positive rate, and the vertical coordinate is the TPR, which is the true positive rate. The AUC calculation formula is as follows:

$$AUC = \frac{\sum_{i \in positiveclass} rank_i - \frac{M \times (M+1)}{2}}{M \times N}, \qquad (10)$$

where $\sum_{i \in positiveclass} rank_i$ refers to the summation of each positive sample serial number located in the position after the subtraction operation (according to the probability of the score from the small to reach the ranking, ranked in the rank position); M and N are the number of positive and negative samples, respectively. MAE and MSE represent the average absolute error and mean square error, respectively, and when both two-evaluation metrics are smaller, it represents the better performance of the algorithm, where the MAE is calculated as follows:

$$MAE = \frac{1}{n} \sum_{i=1}^{n} |\hat{y}_i - y_i|, \qquad (11)$$

The MSE calculation formula is as follows:

$$MSE = \frac{1}{n} \sum_{i=1}^{n} (\hat{y}_i - y_i)^2, \qquad (12)$$

where $\hat{y}_i$ indicates actual value and y_i indicates the predicted value.

For the object detection algorithm, PRE and RECALL are defined in a different way to that in the above statement. The PRE in the object detection algorithm is the percentage of the ratio of the overlapping area size of Ground Truth and the actual localization result to the area size of the actual localization result. RECALL is the percentage of the ratio of the overlapping area size of Ground Truth and the actual localization result to the area size of Ground Truth. The calculation process is shown in Figure 10.

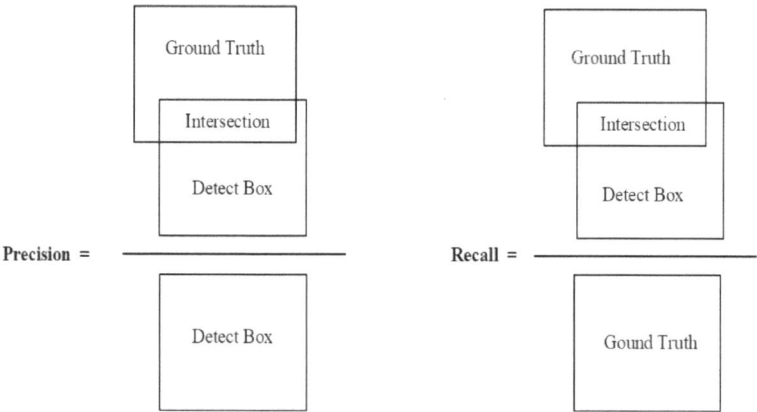

Figure 10. Calculation of PRE and RECALL in the object detection algorithm.

In addition, the calculation of MAE and MSE is slightly different from the above description, but the meaning is the same, so the MAE and MSE calculation formulae are as follows:

$$MAE = \frac{1}{W \times H} \sum_{i=1}^{W \times H} |S(i) - GT_s(i)|, \tag{13}$$

$$MSE = \frac{1}{W \times H} \sum_{i=1}^{W \times H} (S(i) - GT_s(i))^2 \tag{14}$$

where W and H denote the height and width of the localized segmentation area, respectively; S denotes the area of the localized segmentation area; and GT_S denotes the actual defect area. The Dice coefficient, IoU, is also used as an evaluation metric in the object detection algorithm. The Dice coefficient is used to compare the pixel consistency between the predicted segmentation results and their corresponding GT_S, and is calculated as follows:

$$Dice(X, Y) = \frac{2 \times |X \sqcap Y|}{|X| + |Y|}, \tag{15}$$

IoU, which reflects the detection effect of the predicted detection frame and the real detection frame, has a very good characteristic of scale invariance, that is, it is scale insensitive, and the calculation formula is as follows:

$$IoU = \frac{|X \cap Y|}{|X \cup Y|}, \tag{16}$$

where X denotes the segmentation result and Y denotes GT_S. SM [126] can evaluate the structural similarity between the saliency map S and GTs by considering both the values of region perception S_r and object perception S_o, which can be defined as:

$$SM = \alpha \times S_o + (1 - \alpha) \times S_r, \tag{17}$$

where α denotes the balance parameters, which are typically set to 0.5. Finally, PFOM [127] visualizes the boundary quality of the segmentation result, which is commonly used in the edge detection region and can be defined as:

$$PFOM = \frac{1}{\max(N_G, N_S)} \sum_{k=1}^{N_S} \frac{1}{1 + \alpha d_k^2}, \tag{18}$$

where N_G denotes the number of ideal edge points extracted from the GT_S and N_S denotes the actual number of edge points of the segmentation result. α denotes the scaling constants that are typically set to 0.1 or 1/9; d_k denotes the Euclidean distance between the kth true edge point and the detected edge point.

4.2.2. The Efficiency Class Metrics

The efficiency metric is used to evaluate the speed of the relevant algorithm to recognize defects, which can directly reflect whether the algorithm can achieve the ability to identify the product surface defects in real-time on the production line as well as the complexity of the algorithm design.

Commonly used efficiency class metrics include training duration, testing duration, parameters Params, Inference Time, FPS, FLOPs, etc. Among them, training duration and testing duration are the most easily to obtain and intuitive data in the process of algorithm experiments. However, due to the different hardware environment of each work, even if the same method is used, the length of time measured is also different when training and testing on different hardware conditions, which needs to be considered by other efficiency evaluation metrics. Params refers to the number of parameters involved in the calculation of the designed method. This index reflects the amount of memory occupied by the designed method, and its unit is generally represented by M. The Inference Time and FPS are the corresponding evaluation metrics. The former refers to the time spent to detect an image, the unit of which is generally MS; the latter is the number of images that can be detected within one second, also known as the number of frames. FLOPs represents the computing complexity, which describes the complexity of the algorithm and reflects the hardware requirement of the designed algorithm. The unit of FLOPs is usually denoted by B.

5. Challenges and Solutions

This section presents four challenges in the field of steel surface defect detection: the problem of insufficient data samples, the problem of unbalanced data samples, the problem of real-time detection, and the problem of small object detection, and gives appropriate recommendations for solving them.

5.1. The Problem of Insufficient Data Samples

In terms of deep learning algorithms, generally the deeper the level, the better the performance of deep neural networks, but they require a large amount of data for training, otherwise they are prone to overfitting due to the high number of parameters in very deep networks. However, the quality of datasets in the real environment is generally not high and the number of types is not complete, resulting in deep neural networks not being able to give full play to their performance when performing defect detection. To address this problem, there are four effective solutions that can be used in combination with the following. The first is the use of transfer learning methods, which use transfer learning techniques to migrate pre-trained models (e.g., ResNet, VGG, etc.) that have been trained in large datasets to the object detection problem (which usually has a smaller training dataset). Transfer learning relaxes the assumption that the training data must be independent and identically distributed with the test data [128], but it may affect the final algorithm performance because the initial training dataset of the pre-trained model has a large gap with the images for defect detection. Next, unsupervised or weakly supervised methods can be chosen as a way to alleviate the problem of an insufficient and incomplete number of defect samples in the dataset. Detailed descriptions of unsupervised and weakly supervised methods can be found in Sections 3.2.2 and 3.2.3 in this paper. Data augmentation techniques can also be used to expand the dataset to alleviate the problem of insufficient dataset. Data augmentation techniques are very flexible and can be used in the preparation phase of the dataset prior to training or can be performed automatically during training [129]. Common methods of data augmentation include performing image transformation operations such as random cropping, scaling, color shifting, flipping, mirroring, and scaling to pre-process the original image for the purpose of extending the original dataset. Finally, the problem of insufficient data samples is solved by optimizing the network structure such as the proposed GAN network model, which generates the image closest to the test image by continuously optimizing the parameters of the iterative generator G. Then, the defect recognition algorithm is trained based on this, so in the field of surface defect recognition, there are many methods based on GAN networks for image defect recognition methods such as in [130,131].

5.2. The Problem of Unbalanced Data Samples

In deep learning-based defect detection algorithms, when performing model training, the required dataset usually requires the number of sample sets of each category in the dataset to be almost the same. However, the actual situation is that the number of defect-free samples in the dataset is the largest, and the defective samples only account for a small portion, or the number of samples in the defective samples is very unevenly distributed among the various defective categories, and the number of easily collected defective samples accounts for the majority, a situation known as data sample imbalance. This imbalance exists in supervised learning methods, which will directly lead the algorithm to put more attention on the categories with more sufficient samples in the dataset and less recognition ability for the categories with smaller data volume, thus affecting the recognition ability of the algorithm for each type of defect. To address the problem of data sample imbalance, the following methods can be used to solve it. First, at the dataset level, data augmentation, data resampling [132], and GAN networks can be used to change the sample distribution of the training set to a balanced state. Second, at the model level, the attention to small samples can be adjusted by assigning appropriate weights to each sample in the training set, assigning higher weights to a smaller number of sample categories, and lower weights to a larger number of sample categories, and then the objective function can be optimized by increasing the loss term for small classes of misclassified samples in the objective function. In addition, the anomaly detection algorithm can be referred to build a single classifier for anomalies (categories with small sample size) to detect them, thus solving the data sample imbalance problem.

5.3. Real-Time Detection of Problems

In many real industrial scenarios of defect detection tasks, real-time detection is required such as online analysis and online monitoring, which requires a millisecond response speed. Therefore, the ability to perform real-time detection is a very important consideration for the practicality of the algorithm, but the defect recognition algorithm still faces the contradiction of accuracy and speed. For the time being, increasing the detection efficiency of the model can be considered from two aspects: the algorithm and hardware. In terms of algorithms, there is an option to use and develop some lightweight detection algorithms such as SqueezeNet [25], MobileNet [31], and ShuffleNet [32], etc. by trying to improve the recognition speed of the two-stage algorithm or the recognition accuracy of the single-stage algorithm, aiming to achieve the lowest balance between the lowest computational cost and the highest accuracy. In addition, model acceleration can be performed by using optimized convolutional operations or by using distillation, pruning and dropout techniques. On the hardware side, GPUs, FPGAs, DSPs, and Google's TPUs can be used to speed up the model computation.

5.4. The Problem of Small Object Detection

For small object detection, there are two definitions: first, the absolute size is small, and it is generally considered that the object size to be detected is smaller than 32×32 pixels will be recognized as small; the other is the relative size is small, that is, the ratio of the object size to the original image size to be detected is less than 0.1, and it will be recognized as small. The problem of small object detection can be optimized from the following aspects: first, feature fusion technology can be used to fuse deep semantic information into shallow feature maps, while using the rich semantic information of deep features and shallow features to achieve the detection of small objects. Second, one can choose to set the input size of the image by scaling the input image to an appropriate scale and then inputting it into the network for detection. Context information can also be used to establish a connection between the object and its context. Finally, one can also choose to reduce the downsampling rate of the network to reduce the loss of detection objects on the feature map; this reduction in the loss of information is beneficial to the detection of small objects, and common methods are to reduce the pooling layer and use the null convolution.

6. Summary and Future Work

Steel is the core material of the world's industrial production, and the development of steel surface defect recognition technology promotes the high-quality production of steel. Therefore, the research on steel surface defect recognition technology is of great importance worldwide. The various types of defect recognition methods summarized in Section 3 of this paper have been published in the last 10 years, and most of them have been published in the last 5 years and represent the best performance methods in this field. First, this paper summarized the difficulties of steel surface defect recognition, and then counted the surface types of steel and the corresponding surface defect categories. Second, we presented a detailed comparison of the key hardware of the surface defect detection system. This paper presented a systematic classification of steel surface defect recognition algorithms, a literature review for each category, and a summary of the characteristics of each technique. Since there is a lack of large and uniform datasets in the field of steel surface defect recognition, and many methods have different performance evaluation metrics, several commonly used public datasets were summarized and more than 20 performance evaluation metrics were aggregated. Finally, the challenges faced by the current steel surface defect recognition technology are summarized and corresponding solutions are proposed. The goal of this paper was to facilitate the future research efforts of those studying steel surface defect recognition techniques.

In future work, we aim to create a public algorithm performance evaluation platform for steel surface defect recognition, where researchers can make a fair and comprehensive comparison of various steel surface defect recognition algorithms, which will greatly contribute to the development of steel surface defect recognition technology. The platform will provide a complete steel surface defect dataset, a code based on current mainstream public algorithms, a comprehensive set of algorithm performance evaluation metrics, and a software toolkit and interface for researchers to upload new algorithm codes for testing on their own. In addition to building an algorithm evaluation platform, we will design a weakly supervised learning algorithm. We believe that the weakly supervised learning algorithm will become the mainstream algorithm for steel defect recognition in the future because it maintains a similar accuracy and stability of training results to the supervised learning algorithm while significantly reducing the demanding dataset requirements. We will focus our research on weakly supervised learning algorithms in two directions: reducing the algorithm complexity and improving the defect detection efficiency.

Author Contributions: Conceptualization, X.W.; Methodology, J.S.; Validation, J.S. and X.W.; Formal analysis, Y.H.; Investigation, J.S. and X.W.; Resources, K.S.; Data curation, K.S.; Writing—original draft preparation, X.W., J.S., Y.H. and K.S.; Writing—review and editing, X.W., J.S., Y.H. and K.S.; Visualization, J.S. and Y.H.; Supervision, K.S.; Project administration, X.W.; Funding acquisition, X.W. All authors have read and agreed to the published version of the manuscript.

Funding: This research was funded by Liaoning Provincial Department of Education Scientific Research Project, grant number LQGD2020023.

Institutional Review Board Statement: Not applicable.

Informed Consent Statement: Not applicable.

Data Availability Statement: Not applicable.

Conflicts of Interest: The authors declare no conflict of interest.

References

1. Basson, E. World Steel in Figures 2022. Available online: https://worldsteel.org/steel-topics/statistics/world-steel-in-figures-2022/ (accessed on 30 April 2022).
2. Jain, S.; Seth, G.; Paruthi, A.; Soni, U.; Kumar, G. Synthetic data augmentation for surface defect detection and classification using deep learning. *J. Intell. Manuf.* **2020**, *33*, 1007–1020. [CrossRef]
3. He, Y.; Song, K.; Meng, Q.; Yan, Y. An end-to-end steel surface defect detection approach via fusing multiple hierarchical features. *IEEE Trans. Instrum. Meas.* **2019**, *69*, 1493–1504. [CrossRef]
4. Luo, Q.; Fang, X.; Su, J.; Zhou, J.; Zhou, B.; Yang, C.; Liu, L.; Gui, W.; Lu, T. Automated Visual Defect Classification for Flat Steel Surface: A Survey. *IEEE Trans. Instrum. Meas.* **2020**, *69*, 9329–9349. [CrossRef]
5. He, Y.; Wen, X.; Xu, J. A Semi-Supervised Inspection Approach of Textured Surface Defects under Limited Labeled Samples. *Coatings* **2022**, *12*, 1707. [CrossRef]
6. Ma, S.; Song, K.; Niu, M.; Tian, H.; Yan, Y. Cross-scale Fusion and Domain Adversarial Network for Generalizable Rail Surface Defect Segmentation on Unseen Datasets. *J. Intell. Manuf.* **2022**, 1–20. [CrossRef]
7. Wan, C.; Ma, S.; Song, K. TSSTNet: A Two-Stream Swin Transformer Network for Salient Object Detection of No-Service Rail Surface Defects. *Coatings* **2022**, *12*, 1730. [CrossRef]
8. Li, Z.; Tian, X.; Liu, X.; Liu, Y.; Shi, X. A Two-Stage Industrial Defect Detection Framework Based on Improved-YOLOv5 and Optimized-Inception-ResnetV2 Models. *Appl. Sci.* **2022**, *12*, 834. [CrossRef]

9. Song, K.; Wang, J.; Bao, Y.; Huang, L.; Yan, Y. A Novel Visible-Depth-Thermal Image Dataset of Salient Object Detection for Robotic Visual Perception. *IEEE/ASME Trans. Mechatron.* **2022**. [CrossRef]
10. Sun, G.; Huang, D.; Cheng, L.; Jia, J.; Xiong, C.; Zhang, Y. Efficient and Lightweight Framework for Real-Time Ore Image Segmentation Based on Deep Learning. *Minerals* **2022**, *12*, 526. [CrossRef]
11. Neogi, N.; Mohanta, D.K.; Dutta, P.K. Review of vision-based steel surface inspection systems. *EURASIP J. Image Video Process.* **2014**, *2014*, 50. [CrossRef]
12. Verlag Stahleisen GmbH, Germany. Available online: www.stahleisen.de (accessed on 13 June 2022).
13. Viola, P.; Jones, M.J. Robust real-time face detection. *Int. J. Comput. Vis.* **2004**, *57*, 137–154. [CrossRef]
14. Dalal, N.; Triggs, B. Histograms of oriented gradients for human detection. In Proceedings of the 2005 IEEE Computer Society Conference on Computer Vision and Pattern Recognition (CVPR'05), San Diego, CA, USA, 20–25 June 2005; IEEE: Piscatevi, NJ, USA, 2005; 1, pp. 886–893.
15. Felzenszwalb, P.F.; Girshick, R.B.; McAllester, D.; Ramanan, D. Object detection with discriminatively trained part-based models. *IEEE Trans. Pattern Anal. Mach. Intell.* **2010**, *32*, 1627–1645. [CrossRef] [PubMed]
16. Krizhevsky, A.; Sutskever, I.; Hinton, G.E. Imagenet classification with deep convolutional neural networks. *Commun. ACM* **2017**, *60*, 84–90. [CrossRef]
17. Goodfellow, I.; Pouget-Abadie, J.; Mirza, M.; Xu, B.; Warde-Farley, D.; Ozair, S.; Courville, A.; Bengio, Y. Generative adversarial networks. *Commun. ACM* **2020**, *63*, 139–144. [CrossRef]
18. Simonyan, K.; Zisserman, A. Very deep convolutional networks for large-scale image recognition. *arXiv* **2014**, arXiv:preprint,1409,1556.
19. Szegedy, C.; Liu, W.; Jia, Y.; Sermanet, P.; Reed, S.; Anguelov, D.; Rabinovich, A. Going deeper with convolutions. In Proceedings of the IEEE Conference on Computer Vision and Pattern Recognition, Boston, MA, USA, 7–12 June 2015; pp. 1–9.
20. He, K.; Zhang, X.; Ren, S.; Sun, J. Spatial pyramid pooling in deep convolutional networks for visual recognition. *IEEE Trans. Pattern Anal. Mach. Intell.* **2015**, *37*, 1904–1916. [CrossRef]
21. Girshick, R.; Donahue, J.; Darrell, T.; Malik, J. Rich feature hierarchies for accurate object detection and semantic segmentation. In Proceedings of the IEEE Conference on Computer Vision and Pattern Recognition, Columbus, OH, USA, 23–28 June 2014; pp. 580–587.
22. He, K.; Zhang, X.; Ren, S.; Sun, J. Deep residual learning for image recognition. In Proceedings of the IEEE Conference on Computer Vision and Pattern Recognition, Las Vegas, NV, USA, 27–30 June 2016; pp. 770–778.
23. Girshick, R. Fast r-cnn. In Proceedings of the IEEE International Conference on Computer Vision, Santiago, Chile, 7–13 December 2015; pp. 1440–1448.
24. Liu, W.; Anguelov, D.; Erhan, D.; Szegedy, C.; Reed, S.; Fu, C.Y.; Berg, A.C. Ssd: Single shot multibox detector. In *European Conference on Computer Vision*; Springer: Cham, Switzerland, 2016; pp. 21–37.
25. Iandola, F.N.; Han, S.; Moskewicz, M.W.; Ashraf, K.; Dally, W.J.; Keutzer, K. SqueezeNet: AlexNet-level accuracy with 50x fewer parameters and<0.5 MB model size. *arXiv* **2016**, arXiv:preprint. 1602.07360.
26. Ren, S.; He, K.; Girshick, R.; Sun, J. Faster r-cnn: Towards real-time object detection with region proposal networks. *Adv. Neural Inf. Process. Syst.* **2015**, *28*. [CrossRef]
27. Redmon, J.; Divvala, S.; Girshick, R.; Farhadi, A. You only look once: Unified, real-time object detection. In Proceedings of the IEEE Conference on Computer Vision and Pattern Recognition, Las Vegas, NV, USA, 27–30 June 2016; pp. 779–788.
28. Lin, T.Y.; Goyal, P.; Girshick, R.; He, K.; Dollár, P. Focal loss for dense object detection. In Proceedings of the IEEE International Conference on Computer Vision, Venice, Italy, 22–29 October 2017; pp. 2980–2988.
29. Cai, Z.; Vasconcelos, N. Cascade r-cnn: Delving into high quality object detection. In Proceedings of the IEEE Conference on Computer Vision and Pattern Recognition, Los Alamitos, CA, USA, 18–22 June 2018; pp. 6154–6162.
30. Huang, G.; Liu, Z.; Van Der Maaten, L.; Weinberger, K.Q. Densely connected convolutional networks. In Proceedings of the IEEE Conference on Computer Vision and Pattern Recognition, Honolulu, HI, USA, 21–26 July 2017; pp. 4700–4708.
31. Howard, A.G.; Zhu, M.; Chen, B.; Kalenichenko, D.; Wang, W.; Weyand, T.; Adam, H. Mobilenets: Efficient convolutional neural networks for mobile vision applications. *arXiv* **2017**, arXiv:preprint 1704.04861.
32. Zhang, X.; Zhou, X.; Lin, M.; Sun, J. Shufflenet: An extremely efficient convolutional neural network for mobile devices. In Proceedings of the IEEE Conference on Computer Vision and Pattern Recognition, Los Alamitos, CA, USA, 18–22 June 2018; pp. 6848–6856.
33. Duan, K.; Bai, S.; Xie, L.; Qi, H.; Huang, Q.; Tian, Q. Centernet: Keypoint triplets for object detection. In Proceedings of the IEEE/CVF International Conference on Computer Vision, Long Beach, CA, USA, 16–20 June 2019; pp. 6569–6578.
34. Zheng, X.; Zheng, S.; Kong, Y.; Chen, J. Recent advances in surface defect inspection of industrial products using deep learning techniques. *Int. J. Adv. Manuf. Technol.* **2021**, *113*, 35–58. [CrossRef]
35. Wan, X.; Zhang, X.; Liu, L. An Improved VGG19 Transfer Learning Strip Steel Surface Defect Recognition Deep Neural Network Based on Few Samples and Imbalanced Datasets. *Appl. Sci.* **2021**, *11*, 2606. [CrossRef]
36. Wang, S.; Xia, X.; Ye, L.; Yang, B. Automatic detection and classification of steel surface defect using deep convolutional neural networks. *Metals* **2021**, *11*, 388. [CrossRef]
37. Pan, Y.; Zhang, L. Dual attention deep learning network for automatic steel surface defect segmentation. *Comput.-Aided Civ. Infrastruct. Eng.* **2022**, *37*, 1468–1487. [CrossRef]
38. Steger, C.; Ulrich, M.; Wiedemann, C. *Machine Vision Algorithms and Applications*; John Wiley & Sons: Hoboken, NJ, USA, 2018.

39. Hornberg, A. *Handbook of Machine and Computer Vision: The Guide for Developers and User*; John Wiley & Sons: Hoboken, NJ, USA, 2017.
40. Lian, J.; Jia, W.; Zareapoor, M.; Zheng, Y.; Luo, R.; Jain, D.K.; Kumar, N. Deep-learning-based small surface defect detection via an exaggerated local variation-based generative adversarial network. *IEEE Trans. Ind. Inform.* **2019**, *16*, 1343–1351. [CrossRef]
41. Shi, T.; Kong, J.; Wang, X.; Liu, Z.; Zheng, G. Improved Sobel algorithm for defect detection of rail surfaces with enhanced efficiency and accuracy. *J. Cent. South Univ.* **2016**, *23*, 2867–2875. [CrossRef]
42. Lin, H.I.; Wibowo, F.S. Image data assessment approach for deep learning-based metal surface defect-detection systems. *IEEE Access* **2021**, *9*, 47621–47638. [CrossRef]
43. Shreya, S.R.; Priya, C.S.; Rajeshware, G.S. Design of machine vision system for high speed manufacturing environments. In Proceedings of the 2016 IEEE Annual India Conference (INDICON), Bangalore, India, 16–18 December 2016; IEEE: Piscatevi, NJ, USA, 2016; pp. 1–7.
44. Chen, Y.; Ding, Y.; Zhao, F.; Zhang, E.; Wu, Z.; Shao, L. Surface defect detection methods for industrial products: A review. *Appl. Sci.* **2021**, *11*, 7657. [CrossRef]
45. Song, K.; Yan, Y. A noise robust method based on completed local binary patterns for hot-rolled steel strip surface defects. *Appl. Surf. Sci.* **2013**, *285*, 858–864. [CrossRef]
46. Chu, M.; Wang, A.; Gong, R.; Sha, M. Strip steel surface defect recognition based on novel feature extraction and enhanced least squares twin support vector machine. *ISIJ Int.* **2014**, *54*, 1638–1645. [CrossRef]
47. Chu, M.; Gong, R. Invariant feature extraction method based on smoothed local binary pattern for strip steel surface defect. *ISIJ Int.* **2015**, *55*, 1956–1962. [CrossRef]
48. Truong MT, N.; Kim, S. Automatic image thresholding using Otsu's method and entropy weighting scheme for surface defect detection. *Soft Comput.* **2018**, *22*, 4197–4203. [CrossRef]
49. Luo, Q.; Fang, X.; Sun, Y.; Liu, L.; Ai, J.; Yang, C.; Simpson, O. Surface defect classification for hot-rolled steel strips by selectively dominant local binary patterns. *IEEE Access* **2019**, *7*, 23488–23499. [CrossRef]
50. Wang, Y.; Xia, H.; Yuan, X.; Li, L.; Sun, B. Distributed defect recognition on steel surfaces using an improved random forest algorithm with optimal multi-feature-set fusion. *Multimed. Tools Appl.* **2018**, *77*, 16741–16770. [CrossRef]
51. Zhao, J.; Peng, Y.; Yan, Y. Steel surface defect classification based on discriminant manifold regularized local descriptor. *IEEE Access* **2018**, *6*, 71719–71731. [CrossRef]
52. Luo, Q.; Sun, Y.; Li, P.; Simpson, O.; Tian, L.; He, Y. Generalized Completed Local Binary Patterns for Time-Efficient Steel Surface Defect Classification. *IEEE Trans. Instrum. Meas.* **2018**, *63*, 667–679. [CrossRef]
53. Liu, Y.; Xu, K.; Xu, J. An improved MB-LBP defect recognition approach for the surface of steel plates. *Appl. Sci.* **2019**, *9*, 4222. [CrossRef]
54. Xu, K.; Ai, Y.; Wu, X. Application of multi-scale feature extraction to surface defect classification of hot-rolled steels. *Int. J. Miner. Metall. Mater.* **2013**, *20*, 37–41. [CrossRef]
55. Jeon, Y.J.; Choi, D.; Yun, J.P.; Kim, S.W. Detection of periodic defects using dual-light switching lighting method on the surface of thick plates. *ISIJ Int.* **2015**, *55*, 1942–1949. [CrossRef]
56. Xu, K.; Liu, S.; Ai, Y. Application of Shearlet transform to classification of surface defects for metals. *Image Vis. Comput.* **2015**, *35*, 23–30. [CrossRef]
57. Choi, D.; Jeon, Y.J.; Kim, S.H.; Moon, S.; Yun, J.P.; Kim, S.W. Detection of pinholes in steel slabs using Gabor filter combination and morphological features. *ISIJ Int.* **2017**, *57*, 1045–1053. [CrossRef]
58. Ashour, M.W.; Khalid, F.; Abdul Halin, A.; Abdullah, L.N.; Darwish, S.H. Surface Defects Classification of Hot-Rolled Steel Strips Using Multi-directional Shearlet Features. *Arab J. Sci. Eng.* **2019**, *44*, 2925–2932. [CrossRef]
59. Ghorai, S.; Mukherjee, A.; Gangadaran, M.; Dutta, P.K. Automatic defect detection on hot-rolled flat steel products. *IEEE Trans. Instrum. Meas.* **2012**, *62*, 612–621. [CrossRef]
60. Choi, D.C.; Jeon, Y.J.; Lee, S.J.; Yun, J.P.; Kim, S.W. Algorithm for detecting seam cracks in steel plates using a Gabor filter combination method. *Appl. Opt.* **2014**, *53*, 4865–4872. [CrossRef]
61. Liu, X.; Xu, K.; Zhou, D.; Zhou, P. Improved contourlet transform construction and its application to surface defect recognition of metals. *Multidimens. Syst. Signal Process.* **2020**, *31*, 951–964. [CrossRef]
62. Borselli, A.; Colla, V.; Vannucci, M.; Veroli, M. A fuzzy inference system applied to defect detection in flat steel production. In Proceedings of the International Conference on Fuzzy Systems, Barcelona, Spain, 18–23 July 2010; IEEE: Piscatevi, NJ, USA, 2010; pp. 1–6.
63. Liu, M.; Liu, Y.; Hu, H.; Nie, L. Genetic algorithm and mathematical morphology based binarization method for strip steel defect image with non-uniform illumination. *J. Vis. Commun. Image Represent.* **2016**, *37*, 70–77. [CrossRef]
64. Taştimur, C.; Karaköse, M.; Akın, E.; Aydın, I. Rail defect detection with real time image processing technique. In Proceedings of the 2016 IEEE 14th International Conference on Industrial Informatics (INDIN), Poitiers, France, 19–21 July 2016; IEEE: Piscatevi, NJ, USA, 2016; pp. 411–415.
65. Song, K.C.; Hu, S.P.; Yan, Y.H.; Li, J. Surface defect detection method using saliency linear scanning morphology for silicon steel strip under oil pollution interference. *ISIJ Int.* **2014**, *54*, 2598–2607. [CrossRef]
66. Wang, H.; Zhang, J.; Tian, Y.; Chen, H.; Sun, H.; Liu, K. A simple guidance template-based defect detection method for strip steel surfaces. *IEEE Trans. Ind. Inform.* **2018**, *15*, 2798–2809. [CrossRef]

67. Zhou, S.; Wu, S.; Liu, H.; Lu, Y.; Hu, N. Double low-rank and sparse decomposition for surface defect segmentation of steel sheet. *Appl. Sci.* **2018**, *8*, 1628. [CrossRef]
68. Gao, X.; Du, L.; Xie, Y.; Chen, Z.; Zhang, Y.; You, D.; Gao, P.P. Identification of weld defects using magneto-optical imaging. *Int. J. Adv. Manuf. Technol.* **2019**, *105*, 1713–1722. [CrossRef]
69. Xu, K.; Song, M.; Yang, C.; Zhou, P. Application of hidden Markov tree model to on-line detection of surface defects for steel strips. *J. Mech. Eng.* **2013**, *49*, 34. [CrossRef]
70. Wang, J.; Li, Q.; Gan, J.; Yu, H.; Yang, X. Surface defect detection via entity sparsity pursuit with intrinsic priors. *IEEE Trans. Ind. Inform.* **2019**, *16*, 141–150. [CrossRef]
71. Kulkarni, R.; Banoth, E.; Pal, P. Automated surface feature detection using fringe projection: An autoregressive modeling-based approach. *Opt. Lasers Eng.* **2019**, *121*, 506–511. [CrossRef]
72. Ai, Y.; Xu, K. Surface detection of continuous casting slabs based on curvelet transform and kernel locality preserving projections. *J. Iron Steel Res. Int.* **2013**, *20*, 80–86. [CrossRef]
73. Hwang, Y.I.; Seo, M.K.; Oh, H.G.; Choi, N.; Kim, G.; Kim, K.B. Detection and classification of artificial defects on stainless steel plate for a liquefied hydrogen storage vessel using short-time fourier transform of ultrasonic guided waves and linear discriminant analysis. *Appl. Sci.* **2022**, *12*, 6502. [CrossRef]
74. Wang, J.; Fu, P.; Gao, R.X. Machine vision intelligence for product defect inspection based on deep learning and Hough transform. *J. Manuf. Syst.* **2019**, *51*, 52–60. [CrossRef]
75. Hu, M.K. Visual pattern recognition by moment invariants. *IRE Trans. Inf. Theory* **1962**, *8*, 179–187.
76. Hu, H.; Li, Y.; Liu, M.; Liang, W. Classification of defects in steel strip surface based on multiclass support vector machine. *Multimed. Tools Appl.* **2014**, *69*, 199–216. [CrossRef]
77. Hu, H.; Liu, Y.; Liu, M.; Nie, L. Surface defect classification in large-scale strip steel image collection via hybrid chromosome genetic algorithm. *Neurocomputing* **2016**, *181*, 86–95. [CrossRef]
78. Zhang, H.; Zhu, Q.; Fan, C.; Deng, D. Image quality assessment based on Prewitt magnitude. *AEU-Int. J. Electron. Commun.* **2013**, *67*, 799–803. [CrossRef]
79. Fu, G.; Sun, P.; Zhu, W.; Yang, J.; Cao, Y.; Yang, M.Y.; Cao, Y. A deep-learning-based approach for fast and robust steel surface defects classification. *Opt. Lasers Eng.* **2019**, *121*, 397–405. [CrossRef]
80. He, D.; Xu, K.; Zhou, P. Defect detection of hot rolled steels with a new object detection framework called classification priority network. *Comput. Ind. Eng.* **2019**, *128*, 290–297. [CrossRef]
81. Konovalenko, I.; Maruschak, P.; Brezinová, J.; Viňáš, J.; Brezina, J. Steel surface defect classification using deep residual neural network. *Metals* **2020**, *10*, 846. [CrossRef]
82. Zhang, S.; Zhang, Q.; Gu, J.; Su, L.; Li, K.; Pecht, M. Visual inspection of steel surface defects based on domain adaptation and adaptive convolutional neural network. *Mech. Syst. Signal Process.* **2021**, *153*, 107541. [CrossRef]
83. Feng, X.; Gao, X.; Luo, L. X-SDD: A new benchmark for hot rolled steel strip surface defects detection. *Symmetry* **2021**, *13*, 706. [CrossRef]
84. Zhou, X.; Fang, H.; Fei, X.; Shi, R.; Zhang, J. Edge-Aware Multi-Level Interactive Network for Salient Object Detection of Strip Steel Surface Defects. *IEEE Access* **2021**, *9*, 149465–149476. [CrossRef]
85. Li, Z.; Wu, C.; Han, Q.; Hou, M.; Chen, G.; Weng, T. CASI-Net: A novel and effect steel surface defect classification method based on coordinate attention and self-interaction mechanism. *Mathematics* **2022**, *10*, 963. [CrossRef]
86. Hinton, G.E.; Osindero, S.; Teh, Y.W. A fast learning algorithm for deep belief nets. *Neural Comput.* **2006**, *18*, 1527–1554. [CrossRef]
87. Wang, X.B.; Li, J.; Yao, M.H.; He, W.X. Solar cells surface defects detection based on deep learning. *Pattern Recognit. Artif. Intell.* **2014**, *27*, 517–523.
88. Shen, J.; Chen, P.; Su, L.; Shi, T.; Tang, Z.; Liao, G. X-ray inspection of TSV defects with self-organizing map network and Otsu algorithm. *Microelectron. Reliab.* **2016**, *67*, 129–134. [CrossRef]
89. Liu, K.; Wang, H.; Chen, H.; Qu, E.; Tian, Y.; Sun, H. Steel surface defect detection using a new Haar-Weibull-variance model in unsupervised manner. *IEEE Trans. Instrum. Meas.* **2017**, *66*, 2585–2596. [CrossRef]
90. Mei, S.; Yang, H. An unsupervised-learning-based approach for automated defect inspection on textured surfaces. *IEEE Trans. Instrum. Meas.* **2018**, *67*, 1266–1277. [CrossRef]
91. Liu, K.; Li, A.; Wen, X.; Chen, H.; Yang, P. Steel surface defect detection using GAN and one-class classifier. In Proceedings of the 2019 25th International Conference on Automation and Computing (ICAC), Lancaster, UK, 5–7 September 2019; IEEE: Piscatevi, NJ, USA, 2019; pp. 1–6.
92. Youkachen, S.; Ruchanurucks, M.; Phatrapomnant, T.; Kaneko, H. Defect segmentation of hot-rolled steel strip surface by using convolutional auto-encoder and conventional image processing. In Proceedings of the 2019 10th International Conference of Information and Communication Technology for Embedded Systems (IC-ICTES), Bangkok, Thailand, 25–27 March 2019; IEEE: Piscatevi, NJ, USA, 2019; pp. 1–5.
93. Niu, M.; Song, K.; Huang, L.; Wang, Q.; Yan, Y.; Meng, Q. Unsupervised saliency detection of rail surface defects using stereoscopic images. *IEEE Trans. Ind. Inform.* **2021**, *17*, 2271–2281. [CrossRef]
94. Zhou, Z.H. A brief introduction to weakly supervised learning. *Natl. Sci. Rev.* **2018**, *5*, 44–53. [CrossRef]
95. Di, H.; Ke, X.; Peng, Z.; Dongdong, Z. Surface defect classification of steels with a new semi-supervised learning method. *Opt. Lasers Eng.* **2019**, *117*, 40–48. [CrossRef]

96. Berthelot, D.; Carlini, N.; Goodfellow, I.; Papernot, N.; Oliver, A.; Raffel, C.A. Mixmatch: A holistic approach to semi-supervised learning. *Adv. Neural Inf. Process. Syst.* **2019**, *32*.
97. He, Y.; Song, K.; Dong, H.; Yan, Y. Semi-supervised defect classification of steel surface based on multi-training and generative adversarial network. *Opt. Lasers Eng.* **2019**, *122*, 294–302. [CrossRef]
98. Yun, J.P.; Shin, W.C.; Koo, G.; Kim, M.S.; Lee, C.; Lee, S.J. Automated defect inspection system for metal surfaces based on deep learning and data augmentation. *J. Manuf. Syst.* **2020**, *55*, 317–324. [CrossRef]
99. Božič, J.; Tabernik, D.; Skočaj, D. Mixed supervision for surface-defect detection: From weakly to fully supervised learning. *Comput. Ind.* **2021**, *129*, 103459. [CrossRef]
100. Tabernik, D.; Šela, S.; Skvarč, J.; Skočaj, D. Segmentation-based deep-learning approach for surface-defect detection. *J. Intell. Manuf.* **2020**, *31*, 759–776. [CrossRef]
101. Božič, J.; Tabernik, D.; Skočaj, D. End-to-end training of a two-stage neural network for defect detection. In Proceedings of the 2020 25th International Conference on Pattern Recognition (ICPR), Milan, Italy, 10–15 January 2021; IEEE: Piscatevi, NJ, USA, 2021; pp. 5619–5626.
102. Zhang, J.; Su, H.; Zou, W.; Gong, X.; Zhang, Z.; Shen, F. CADN: A weakly supervised learning-based category-aware object detection network for surface defect detection. *Pattern Recognit.* **2021**, *109*, 107571. [CrossRef]
103. Li, J.; Su, Z.; Geng, J.; Yin, Y. Real-time detection of steel strip surface defects based on improved yolo detection network. *IFAC-PapersOnLine* **2018**, *51*, 76–81. [CrossRef]
104. Yang, J.; Li, S.; Wang, Z.; Yang, G. Real-time tiny part defect detection system in manufacturing using deep learning. *IEEE Access* **2019**, *7*, 89278–89291. [CrossRef]
105. Cheng, X.; Yu, J. RetinaNet with difference channel attention and adaptively spatial feature fusion for steel surface defect detection. *IEEE Trans. Instrum. Meas.* **2020**, *70*, 1–11. [CrossRef]
106. Kou, X.; Liu, S.; Cheng, K.; Qian, Y. Development of a YOLO-V3-based model for detecting defects on steel strip surface. *Measurement* **2021**, *182*, 109454. [CrossRef]
107. Chen, X.; Lv, J.; Fang, Y.; Du, S. Online Detection of Surface Defects Based on Improved YOLOV3. *Sensors* **2022**, *22*, 817. [CrossRef] [PubMed]
108. Tian, R.; Jia, M. DCC-CenterNet: A rapid detection method for steel surface defects. *Measurement* **2022**, *187*, 110211. [CrossRef]
109. Wei, R.; Song, Y.; Zhang, Y. Enhanced faster region convolutional neural networks for steel surface defect detection. *ISIJ Int.* **2020**, *60*, 539–545. [CrossRef]
110. Zhao, W.; Chen, F.; Huang, H.; Li, D.; Cheng, W. A new steel defect detection algorithm based on deep learning. *Comput. Intell. Neurosci.* **2021**, *2021*. [CrossRef]
111. Natarajan, V.; Hung, T.Y.; Vaikundam, S.; Chia, L.T. Convolutional networks for voting-based anomaly classification in metal surface inspection. In Proceedings of the 2017 IEEE International Conference on Industrial Technology (ICIT), Toronto, ON, Canada, 22–25 March 2017; IEEE: Piscatevi, NJ, USA, 2017; pp. 986–991.
112. Ren, R.; Hung, T.; Tan, K.C. A generic deep-learning-based approach for automated surface inspection. *IEEE Trans. Cybern.* **2017**, *48*, 929–940. [CrossRef]
113. Liu, Y.; Xu, K.; Xu, J. Periodic surface defect detection in steel plates based on deep learning. *Appl. Sci.* **2019**, *9*, 3127. [CrossRef]
114. Song, L.; Lin, W.; Yang, Y.-G.; Zhu, X.; Guo, Q.; Xi, J. Weak micro-scratch detection based on deep convolutional neural network. *IEEE Access* **2019**, *7*, 27547–27554. [CrossRef]
115. He, D.; Xu, K.; Wang, D. Design of multi-scale receptive field convolutional neural network for surface inspection of hot rolled steels. *Image Vis. Comput.* **2019**, *89*, 12–20. [CrossRef]
116. Yang, H.; Chen, Y.; Song, K.; Yin, Z. Multiscale feature-clustering-based fully convolutional autoencoder for fast accurate visual inspection of texture surface defects. *IEEE Trans. Autom. Sci. Eng.* **2019**, *16*, 1450–1467. [CrossRef]
117. Zhou, F.; Liu, G.; Ni, H.; Ren, F. A generic automated surface defect detection based on a bilinear model. *Appl. Sci.* **2019**, *9*, 3159. [CrossRef]
118. Zhang, J.; Kang, X.; Ni, H.; Ren, F. Surface defect detection of steel strips based on classification priority YOLOv3-dense network. *Ironmak. Steelmak.* **2021**, *48*, 547–558. [CrossRef]
119. Lin, C.Y.; Chen, C.H.; Yang, C.Y.; Akhyar, F.; Hsu, C.Y.; Ng, H.F. Cascading convolutional neural network for steel surface defect detection. In Proceedings of the International Conference on Applied Human Factors and Ergonomics, Washington, DC, USA, 24–28 July 2019; Springer: Cham, Switzerland, 2019; pp. 202–212.
120. Song, K.; Yan, Y. Micro surface defect detection method for silicon steel strip based on saliency convex active contour model. *Math. Probl. Eng.* **2013**, *2013*, 1–13. [CrossRef]
121. Buscema, M.; Terzi, S.; Tastle, W. A new meta-classifier. In Proceedings of the 2010 Annual Meeting of the North American Fuzzy Information Processing Society (NAFIPS), Toronto, ON, Canada, July 2010; pp. 1–7.
122. Lv, X.; Duan, F.; Jiang, J.J.; Fu, X.; Gan, L. Deep metallic surface defect detection: The new benchmark and detection network. *Sensors* **2020**, *20*, 1562. [CrossRef]
123. Gan, J.; Li, Q.; Wang, J.; Yu, H. A hierarchical extractor-based visual rail surface inspection system. *IEEE Sens. J.* **2017**, *17*, 7935–7944. [CrossRef]
124. DAGM 2007 Datasets. Available online: https://hci.iwr.uni-heidelberg.de/node/3616 (accessed on 25 February 2021).

125. Kylberg, G. The Kylberg Texture Dataset, V. 1.0. In *Technical Report 35*; Centre Image Anal., Swedish University of Agricultural Sciences: Uppsala, Sweden, 2011.
126. Fan, D.P.; Cheng, M.M.; Liu, Y.; Li, T.; Borji, A. Structure-measure: A new way to evaluate foreground maps. In Proceedings of the IEEE International Conference on Computer Vision, Venice, Italy, 22–29 October 2017; pp. 4548–4557.
127. Abdou, I.E.; Pratt, W.K. Quantitative design and evaluation of enhancement/thresholding edge detectors. *Proc. IEEE* **1979**, *67*, 753–763. [CrossRef]
128. Tan, C.; Sun, F.; Kong, T.; Yang, C.; Liu, C. A survey on deep transfer learning. In *International Conference on Artificial Neural Networks*; Springer: Cham, Switzerland, 2018; pp. 270–279.
129. Mujeeb, A.; Dai, W.; Erdt, M.; Sourin, A. Unsupervised surface defect detection using deep autoencoders and data augmentation. In Proceedings of the 2018 International Conference on Cyberworlds (CW), Singapore, 3–5 October 2018; IEEE: Piscatevi, NJ, USA, 2018; pp. 391–398.
130. Niu, S.; Li, B.; Wang, X.; Lin, H. Defect image sample generation with GAN for improving defect recognition. *IEEE Trans. Autom. Sci. Eng.* **2020**, *17*, 1611–1622. [CrossRef]
131. Schlegl, T.; Seebck, P.; Waldstein, S.M.; Langs, G.; Schmidt-Erfurth, U. f-AnoGAN: Fast unsupervised anomaly detection with generative adversarial networks. *Med. Image Anal.* **2019**, *54*, 30–44. [CrossRef] [PubMed]
132. Li, M.; Xiong, A.; Wang, L.; Deng, S.; Ye, J. ACO Resampling: Enhancing the performance of oversampling methods for class imbalance classification. *Knowl.-Based Syst.* **2020**, *196*, 105818. [CrossRef]

Disclaimer/Publisher's Note: The statements, opinions and data contained in all publications are solely those of the individual author(s) and contributor(s) and not of MDPI and/or the editor(s). MDPI and/or the editor(s) disclaim responsibility for any injury to people or property resulting from any ideas, methods, instructions or products referred to in the content.

Article

Detection of Complex Features of Car Body-in-White under Limited Number of Samples Using Self-Supervised Learning

Chuang Liu, Kang Su, Long Yang *, Jie Li and Jingbo Guo

State Key Laboratory of Mechanical Behavior and System Safety of Traffic Engineering Structures, Shijiazhuang Tiedao University, Shijiazhuang 051130, China; liuc@stdu.cn.edu (C.L.); sukang2017@163.com (K.S.); lijie@stdu.edu.cn (J.L.); guojingbo66@163.com (J.G.)
* Correspondence: yanglong233@163.com; Tel.: +86-136-2334-8742

Abstract: The measurement and monitoring of the dimensional characteristics of the body-in-white is an important part of the automobile manufacturing process. The process of using key point regression technology to perform online detection of complex features on body-in-white currently faces a bottleneck problem, namely limited training samples. Under the condition that the number of labeled normal map samples is limited, this paper proposes a framework for domain-independent self-supervised learning using a large number of original images. Under this framework, a self-supervised pre-order task is designed, which uses a large number of easily accessible unlabeled original images for characterization learning as well as a domain discriminator to conduct adversarial training on the feature extractor, so that the extracted representation is domain-independent. Finally, in the key point regression task of five different complex features, a series of comparative experiments were carried out between the method in this paper and benchmark methods such as supervised learning, conventional self-supervised learning, and domain-related self-supervised learning. The results show that the method proposed in this paper has achieved significant performance advantages. In the principal component analysis of extracting features, the representation extracted by the method in this paper does not show obvious domain information.

Keywords: body-in-white; complex features; detection; self-supervised learning; the training sample

1. Introduction

The rapid development of the automobile industry makes it occupy an important position in the global manufacturing industry and also makes automobile manufacturers put forward higher requirements for product quality. Accurately measuring and monitoring all kinds of dimensional features of an automobile body-in-white, especially complex features, is an important means to improve the quality control of automobile manufacturing. Figure 1 shows the comparison between simple features and complex features. Conventional 3D measurement technology is based on surface point cloud registration, which can better capture the surface information of complex features, but it is difficult to perform high-precision 3D reconstruction of these complex features [1–3]. This problem can be solved by using the keypoint regression method [4,5]. However, the core algorithm of this method is to regress keypoint coordinates from a two-dimensional normal map, and its training process requires a large number of normal maps with keypoint location labels. For industrial inspection and keypoint regression using surface normal maps, there is currently no database available in the industry. The acquisition of the normal map requires taking a large number of complex feature photos under multiple light sources and using photometric stereo technology to reconstruct the surface normal vector. Key point location calibration needs to go through relative position measurement of key points, camera calibration and coordinate transformation, which leads to a considerable cost to obtain the training data. Therefore, the question concerning how to complete the key point regression task based on normal map under the condition of limited number of labeled samples has become a challenging topic in complex feature detection.

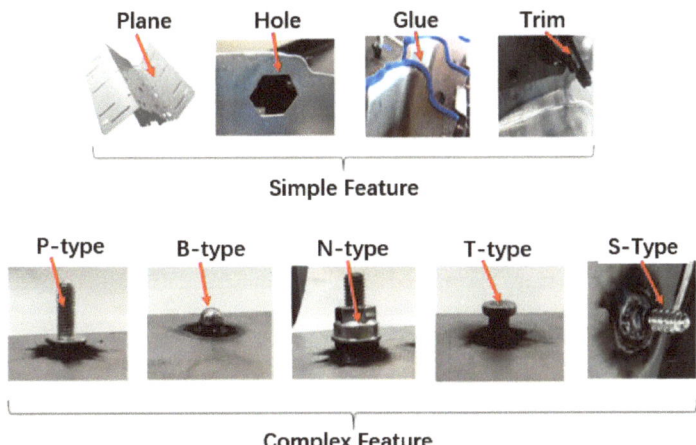

Figure 1. Comparison of simple features and complex features.

At present, the research on the key point regression problem is generally based on the deep learning technology of convolutional neural network, most of which are supervised learning [6]. The training process of supervised learning often requires a large amount of labeled data. In order to reduce the dependence of the keypoint regression method on the labeled training data, scholars try to use unlabeled data for self-supervised learning [7]. Inspired by self-supervised learning, Doersch et al. [8] proposed to learn high-level image representations by using predicting the relative positions of image patches as a pre-order task of self-supervised learning. This work has spawned research on patch-based visual representations for self-supervised learning, where Noroozi et al.'s model predicts the arrangement of "jigsaw puzzles" created from complete pictures [9]. Others generate well-designed image-level classification tasks in contrast to chunking-based methods. Caron et al. [10] used image clustering to create class labels, with classification tasks as pre-tasks. There are other pre-tasks for these methods such as image inpainting [11], image colorization [12], motion segmentation prediction [13] and spatial output tasks with high density. However, none of the above algorithms are applicable to the normal map-based keypoint regression problem.

Many scholars have studied the use of original images for self-supervised learning, and downstream tasks are also performed on the original images, but the representation of complex features under the original images is not obvious. The key point regression problem based on the original images is difficult to achieve satisfactory accuracy [14]. Research [15] shows that normal maps of complex features contain richer information than raw images. If self-supervision is performed in the original image and downstream tasks are performed in the normal map, the representation extracted by self-supervised learning will contain domain information, that is, the domain gap between the two images makes the representation learned by self-supervision in the poor performance in downstream tasks [16].

To address this challenge, this study proposed a highly data-efficient domain-invariant-self-supervised learning method (DISS). The model used by this method consists of two parts: a feature extractor and a regressor. Feature extraction used a jig-saw-based self-supervised pre-task (also translated as agent task) for training with both labeled and unlabeled data. In order to eliminate the domain distance between the original image and the surface normal map, this algorithm used a discriminator to assist training, and reduced the "KL" divergence between the probability distribution of the original image and the surface normal map through adversarial training. The training of the regressor was performed entirely using labeled data.

This paper was divided into four parts. Part 1 was the introduction. Part 2 elaborated the framework of domain-independent self-supervised learning, including training data and problem description, self-supervised pretext-task training, network architecture, and training process. Part 3 was the experimental results and analysis, including the acquisition of data sets, complex feature key point regression experiments, and principal component analysis of extracted features. Part 4 was the conclusion.

2. Domain-Independent Self-Supervised Learning Framework

2.1. Training Data and Problem Description

Given a small amount of labeled normal map $D_t = \{(x_i^t, y_i^t)\}_{i=1}^{m_t}$, in which m_t represented the number of surface normal map, x_i^t represented the surface normal map, and y_i^t represented the corresponding label. Given a large amount of unlabeled raw image $D_s = \{(x_i^s)\}_{i=1}^{m_s}$, in which m_s represented the amount of raw image, x_i^s represents the raw image. According to the assumptions of the question in this paper, there was $m_s \gg m_t$. Given a test set $D_v = \{(x_i^v, y_i^v)\}_{i=1}^{m_v}$, in which m_v represented the number of test data, x_i^v represented the surface normal map, and y_i^v represented the corresponding label. The goal of DISS was to use only D_t and D_s for training and to minimize the error on the test set D_v. The DISS training process was divided into two stages: training on the pre-task of the feature extractor, and training on the downstream task of the regressor.

2.2. Pre-Task Training

In self-supervised learning, pseudo-labels are usually obtained from unlabeled data to form pre-tasks. The design of the pre-task must meet two conditions: it can make the neural network converge, and at the same time, it must learn useful knowledge. Typical pre-tasks include: jig-saw puzzle, image rotation, image colorization, image completion, etc. In this problem, the jig-saw puzzle works best. Since the surface normal map was obtained from the original image through the photometric stereo algorithm, each normal map in the dataset D_t corresponded to at least one original image, and these original images were formed into a dataset $D_r = \{(x_i^r, y_i^r)\}_{i=1}^{m_t}$. Given training dataset D_t, D_r and D_s, in which x_i^s, x_i^r and x_i^t were divided into m_p jig-saw pieces. The nine pictures were randomly shuffled in order, and the shuffled different sequences were one-hot encoded to form pseudo-labels y_i^{s*}, y_i^{r*} and y_i^{t*}. They formed new data sets $D_t^* = \left\{(x_{i,j}^t, y_i^{*t})_{j=1}^{m_p}\right\}_{i=1}^{m_t}$, $D_r^* = \left\{(x_{i,j}^r, y_i^{*r})_{j=1}^{m_p}\right\}_{i=1}^{m_t}$ and $D_s^* = \left\{(x_{i,j}^s, y_i^{*s})_{j=1}^{m_p}\right\}_{t=1}^{m_s}$, which were used to train the pre-task. Figure 2 simplified the acquisition of the pre-order task using a four-piece puzzle. If the feature extractor can correctly distinguish the order of the puzzles, the features extracted by the extractor must also contain some kind of representation of the relationship between different parts of the object.

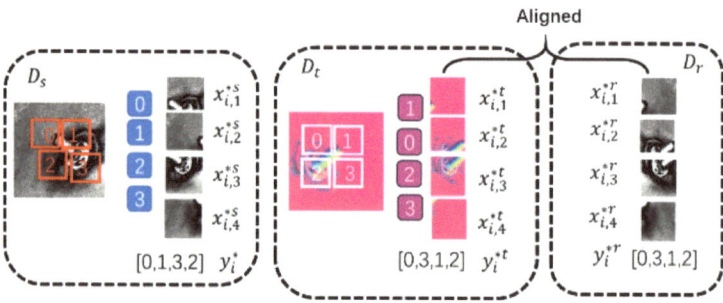

Figure 2. Schematic diagram of jigsaw pre-order task.

2.3. Network Architecture

The neural network architecture was shown in Figure 3. The network was divided into four parts: feature extractor E, sorter S, domain discriminator C and regressor R. As shown in the Figure 3, the training of DISS was divided into three stages. In the first stage of pre-task training, the features extracted by the feature extractor were sent to the sorter and the domain discriminator respectively. The target features extracted by the feature extractor had domain-independent variability through adversarial learning. During the regressor training process in the second stage, the parameters in the feature extractor were frozen, and only the parameters of the regressor were updated in the network. In the final fine-tuning stage, the parameters in the feature extractor and regressor were trained simultaneously.

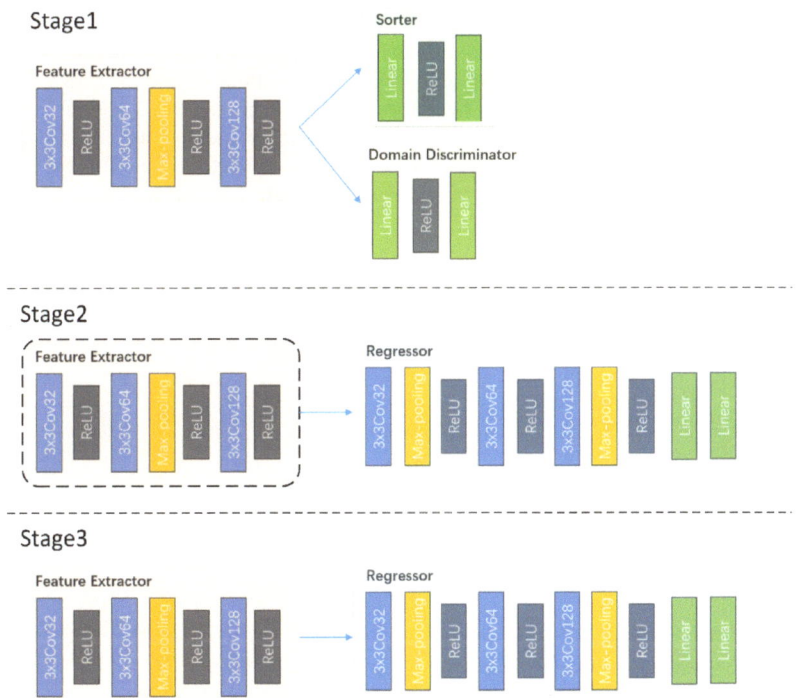

Figure 3. Network architecture.

The role of the feature extractor was to extract representations from the input that was useful for downstream tasks. The feature extractor consisted of 3 convolutional layers and a maxpooling layer, where the convolutional layers all used small convolution kernels.

The sorter was used for the pre-task. It sorted the features extracted by the feature extractor, and extracted the information of the object structure by forcing the feature extractor through learning. The sorter used a fully connected layer and a softmax layer as a classifier to predict the shuffled order of the jig-saw puzzle in the pre-task. The total number of categories of the pre-order task is equal to the factorial of the number of jig-saw puzzle pieces, which will lead to very many classification categories, and too many categories will cause training difficulties. Therefore, a small amount of all possible out-of-order combinations was usually extracted, and then one-hot encoding was performed. In this problem, 50 different out-of-order combinations were taken.

A large number of unlabeled samples in the training dataset came from the original image dataset, whose data distribution had a domain distance from the surface normal map dataset used for testing. The introduction of domain discriminator and adversarial training, on the one hand, was used to discriminate whether the extracted features were from D_t or D_s, on the other hand, it could reduce the domain distance between the features extracted from the original image domain and the surface normal map domain.

The regressor was used to estimate the complex feature keypoint locations through the extracted features, which consisted of three convolutional layers and one fully connected layer.

2.4. Training Process

The training process of DISS was divided into three stages: a pre-task stage, training regressor stage, and fine-tune stage.

In the pre-task stage, the data sets $D_t^* = \left\{ (x_{i,j}^t, y_i^{*t})_{j=1}^{m_p} \right\}_{i=1}^{m_t}$, $D_r^* = \left\{ (x_{i,j}^r, y_i^{*r})_{j=1}^{m_p} \right\}_{i=1}^{m_t}$ and $D_s^* = \left\{ (x_{i,j}^s, y_i^{*s})_{j=1}^{m_p} \right\}_{i=1}^{m_s}$ were used to train the feature extractor, domain discriminator and sorter, and the process was shown in Figure 4. To simplify the description, the example in Figure 4 was the case of a four-piece puzzle. The surface normal map x_i^t was decomposed into four small pictures, and four one-dimensional feature vectors were obtained through the feature extractor E. The four vectors were shuffled in order y_i^{*t} and concatenated into one dimension to form the representation:

$$f_i^{*t} = Cat(\left\{ E(x_{i,j}^t) \right\}_{j=1}^{m_p}) \tag{1}$$

where $Cat(\ )$ represented the vector concatenation operation. Did the same with D_r^* and D_s^* to get the representation:

$$f_i^{*s} = Cat(\left\{ E(x_{k,j}^s) \right\}_{j=1}^{m_p}), f_i^{*s} = Cat(\left\{ E(x_{k,j}^s) \right\}_{j=1}^{m_p}) \tag{2}$$

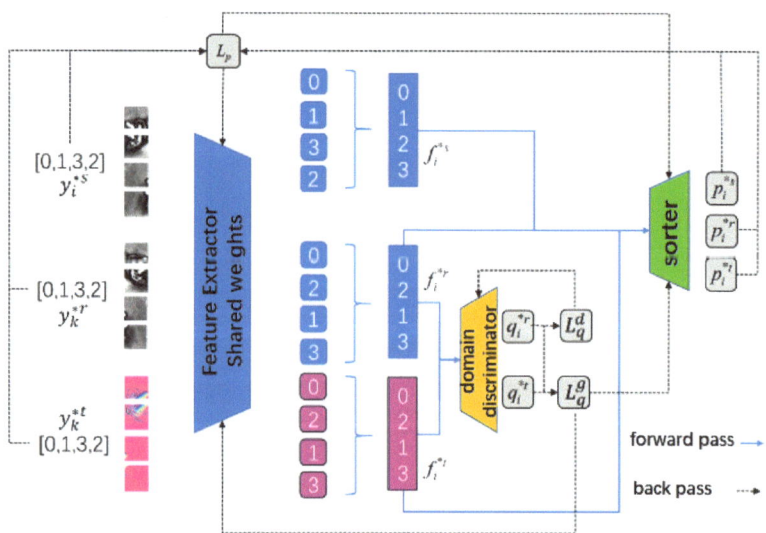

Figure 4. Data flow in the pre-task stage.

After the features were extracted, the sorter needed to sort them. The representation f_i^{*t} was input to the sorter:

$$p_i^{*t} = S(f_i^{*t}), p_i^{*s} = S(f_i^{*s}), p_i^{*r} = S(f_i^{*r}) \tag{3}$$

The domain discriminator needed to determine which domain (original image or surface normal map) it came from, then the features were input into the domain discriminator to get:

$$q_i^{*r} = C(f_i^{*r}), q_i^{*t} = C(f_i^{*t}) \tag{4}$$

The ranking loss function L_p was defined as:

$$L_p = \sum_{i=1}^{m_s} Ce(p_i^{*r}, y_i^{*r}) + Ce(p_i^{*t}, y_i^{*t}) + Ce(p_i^{*s}, y_i^{*s}) \tag{5}$$

where $Ce(\ )$ was the cross entropy loss function.

Defined the discriminative domain loss function L_q^d:

$$L_q^d = \sum_{i=1}^{m_s} -\log(1 - q_i^{*r}) - \log(q_i^{*t}) \tag{6}$$

Defined the extractor domain loss function L_q^g:

$$L_q^g = \sum_{i=1}^{m_s} -\log(1 - q_i^{*t}) \tag{7}$$

When training the pre-order task, the feature extractor and the sorter were trained simultaneously, and the domain discriminator was trained alternately with the other two neural networks. The pseudocode of an epoch training process was shown in Algorithm 1. Among them, λ was the learning rate, θ_E, θ_C, and θ_S were the neural network parameters of the feature extractor, the domain discriminator, and the domain sorter, respectively.

Algorithm 1 Pre-task training algorithm

Input: $D_t^* = \left[\left(x_{ij}^t, y_j^{*t}\right)_{j=1}^{m_p}\right]_{i=1}^{m_t}$; $D_r^* = \left[\left(x_{ij}^r, y_j^{*r}\right)_{j=1}^{m_p}\right]_{i=1}^{m_t}$; $D_s^* = \left[\left(x_{ij}^s, y_j^{*s}\right)_{j=1}^{m_p}\right]_{i=1}^{m_s}$; λ

Initialize weights $\theta_E, \theta_C, \theta_S$
for $i = 1 \rightarrow m_s$ do
    Extract features $f_i^{*t}, f_i^{*s}, f_i^{*r}$ according to formula (1) and (2)
    if $mod(i, 10) < 5$ then
        Calculate q_i^{*r}, q_i^{*t} according to formula (4)
        Calculate the discriminative domain loss function L_q^d according to formula (6)
        Calculate the extractor domain loss function L_q^g according to formula (7)
        Update $\theta_C \leftarrow \theta_C - \lambda \Delta L_q^d$
    else
        Calculate the sort prediction $p_i^{*r}, p_j^{*t}, p_j^{*s}$ according to formula (3)
        Calculate the ranking loss function L_p according to formula (5)
        Update $\theta_E \leftarrow \theta_E - \lambda \Delta (L_q^g + L_p)$
        Update $\theta_S \leftarrow \theta_S - \lambda \Delta L_p$
    end if
end for

The training of downstream tasks was divided into two stages: training the regressor and fine-tuning the model.

Regression training used only the dataset $D_t = \{(x_i^t, y_i^t)\}_{i=1}^{m_t}$. The normal map in the dataset was input into the pre-task trained extractor to get the representation:

$$f_i = E(x_i^t) \tag{8}$$

The features were fed into the regressor to get the predicted coordinates of the keypoints:

$$\hat{y}_i = R(f_i) \tag{9}$$

The regression loss was calculated:

$$L_r = \sum_{i=1}^{m_t} MSE(\hat{y}_i, y_i^t) \qquad (10)$$

Training process contained a total of δ epochs. The first δ_f epochs were the training regressor stage, which froze the model parameters of the feature extractor. After the δ_fth epoch, the fine-tuning stage was entered, and the feature extractor and regressor were trained at the same time. The pseudocode of the training process was shown in Algorithm 2.

Algorithm 2 Regression task training algorithm

Input: $D_t^* = [(x_i^t, y_i^t)]_{i=1}^{m_t}, \lambda_r, \tau, \delta, \delta_f$
Initialize weights θ_R, θ_E
for j = 1 → δ do
   for i = 1 → m_t do
      Extract features f_i^t according to formula (8)
      Extract features $\hat{y}_i$ according to formula (9)
      Calculate L_r
   end for
   if $j < \delta_f$ then
      Update $\theta_R \leftarrow \theta_R - \lambda_r \Delta L_r$
   else
      Update $\theta_R \leftarrow \theta_R - \lambda_r \Delta L_r$
      Update $\theta_E \leftarrow \theta_E - \lambda_r \Delta L_r$
   end if
   if $mod(j, \tau) == 0$ then
      $\lambda_r \leftarrow \lambda_r / 2$
   end if
end for

---

3. Experimental Results and Analysis

3.1. Acquisition of Datasets

In order to evaluate the performance of the DISS framework proposed in this study, a series of experiments were carried out in this paper. The experiment used the complex feature database proposed by Liu et al. [15], in which the labels of the original image data were completely hidden, and the labels of the normal map part were partly hidden, so as to simulate the situation of missing data labels. Figure 5 described the process of obtaining the four datasets used in this paper.

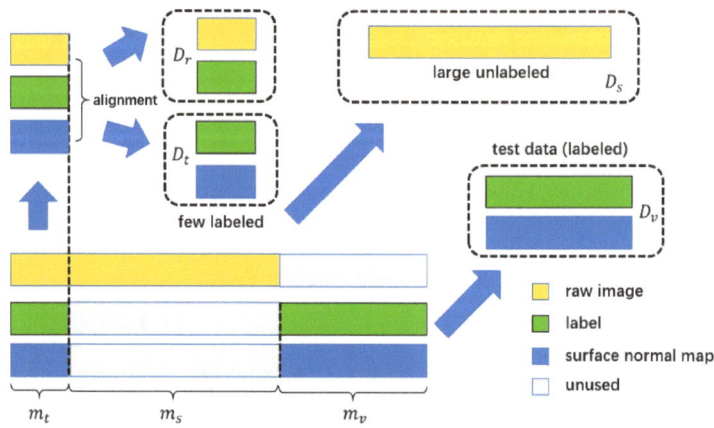

Figure 5. The datasets used in the experiment.

Liu's dataset contains raw images, surface normal maps, labels, and was referred to here as the original dataset. The m_t sets of data containing the surface normal map and the label were taken from the original dataset to form the normal map with labels dataset D_t. The m_r sets of data containing the original image and label were taken from this dataset to form the original image with labels dataset D_r. Note that the D_t and D_r were aligned, so they can use the same label data.

The m_s sets of raw image data were taken from the original dataset to form the original image dataset D_s. According to the hypothesis of the research question, there was $m_t \ll m_s$. Finally, the m_v sets of data containing the surface normal map and the label were taken from the original dataset to verify the accuracy of the algorithm.

3.2. Complex Feature Keypoint Regression Experiments

The methods proposed in this paper were compared with the benchmark methods SL (supervised learning), RSS (raw images self-supervised), and NRSS (normal maps and raw images self-supervised). The SL adopted the direct regression method, which only used labeled data for training. The RSS employed a Jig-saw-based self-supervised method, using unlabeled data for pre-task training. The NRSS method did not use adversarial training with a domain discriminator. The benchmark methods used for comparison were described in detail in Table 1.

Table 1. Comparison of different methods in the experiment.

Method	Peculiarity			
	Use the Original Image	Surface Normal Diagram	Self-Supervision	Confrontation Training
SL	not	Yes	not	not
RSS	Yes	Yes	not	not
NRSS	Yes	Yes	Yes	not
DISS	Yes	Yes	Yes	Yes

Here is a brief introduction to the difference between supervised learning and self-supervised learning. Training samples for supervised learning must be labeled sample data. Supervised learning is to train a model (learn a function) from a given labeled training data sets. When new test data is input, the result can be predicted according to the function. Self-supervised learning mainly uses proxy tasks to mine its own supervision information from large-scale unlabeled data, and trains the network through the constructed supervision information, so that it can learn valuable representations for downstream tasks.

Define the ratio of labeled data to unlabeled data,

$$\alpha = m_t/m_s \qquad (11)$$

The ratio of unlabeled data to labeled data would affect how much the performance improvement brought by self-supervised method would be. This parameter was proposed for two reasons: First, it was to clarify the number of labeled data. Second, we could analyze the impact of changes in the number of labeled samples on DISS by changing the value of this parameter.

α = 5% was adopted in Experiment 1, and the results were shown in Table 2.

In the experiments, the DISS proposed in this paper had the lowest average error. For the four complex features of N, P, S, and T types, DISS has the lowest error. For complex features of B type, the error of the DISS method was also very close to the lowest value. The experimental results fully reflected the advantages of DISS. Through self-supervised training, DISS was able to learn useful information about the geometric features of complex features in the pre-task using a large amount of unlabeled data, thereby gaining an advantage in the downstream task. Compared with RSS and NRSS, which also used self-supervised learning, DISS still had significant advantages. This was due to the fact that although RSS used self-supervised learning, the input data for the pre-task and downstream tasks came from different domains, and this inter-domain bias resulted in that self-supervision could only improve the final performance to a limited extent. Comparing RSS and NRSS, it could be found that without using the domain discriminator and adversarial training, adding the normal map to training dataset did not improve the performance of the network on downstream tasks. This is due to the fact that although the discriminator was able to rank the input data from two different domains correctly in pre-tasks, there was no guarantee that the same representations were extracted from the input data from the two different domains. The structure information of complex feature learned by the feature extractor in the pre-task was mixed with different information from the two domains. This made it impossible to ensure that useful information can be extracted from input data from a single domain in downstream tasks.

Table 2. The average error of the proposed method and the three benchmark methods on five complex features in Experiment 1.

The Characteristic Type	Feature Recognition	Average Error (in Pixels)			
		DISS (Our Method)	SL	RSS	NRSS
N-type features	nut bolt	9.22	11.88	10.82	11.12
P-type features	flat bolt	9.87	12.42	10.61	10.79
S-type features	standard bolt	8.23	9.33	8.88	8.94
T-type features	t-bolt	8.55	8.87	8.75	8.76
Type B features	Ball-stud	7.43	7.42	7.44	7.56
average	-	8.66	9.98	9.30	9.43

Compared with the SL method, DISS had more obvious advantages in N-type, P-type and S-type features. The reason was that these three kinds of complex features were relatively large in size and had rich texture and geometric features. In contrast, T-type and B-type features were smaller in size and had centrosymmetric shapes. In the pre-task, complex features with larger volume and richer surface information are easier to distinguish, while complex features with small volume and centrosymmetric shape are difficult to distinguish.

Experiment 2 was carried out when α takes three different values. Table 3 presented the results of Experiment 2. For the sake of brevity, only the average error of all kinds of complex features was shown here.

Table 3. The mean value of errors under different values of α.

α Value	Average Error (in Pixels)			
	DISS (Text Method)	SL	RSS	NRSS
5%	8.66	9.98	9.30	9.43
10%	7.69	9.07	8.23	8.20
20%	5.88	6.38	6.40	6.48
40%	4.52	4.68	6.62	6.59

The results of experiment 2 showed that as the amount of labeled data increased, the advantage of DISS over SL gradually decreased. This was due to the fact that the fourth stage of DISS had the same effect as the last few epochs of SL. When the number of labeled samples gradually increased, the effect of self-supervision was gradually masked by supervised learning.

3.3. Principal Component Analysis of Extracted Features

In order to further analyze the influence of DISS on the feature extractor, a principal component analysis (primary components analysis, PCA) experiment of extracted features was carried out in this paper. The images x_i in one dataset were imported into the feature extractor to get the representation of all samples, which were then expanded into n-dimensional vectors $\vec{f}_i' \in \mathbb{R}^{n \times 1}$, and all of the extracted features were represented as a matrix:

$$F' = [f_1', f_2', \ldots, f_m'] \tag{12}$$

To normalize the data:

$$f_{i,j} = \frac{f'_{i,j} - \overline{f}_i'}{\sqrt{s_{i,i}}}, i = 1, 2, \ldots, n, j = 1, 2, \ldots, m \tag{13}$$

In which $f_{i,j}$ represented the elements in the normalized matrix F. To calculate its sample correlation matrix as:

$$R = \frac{1}{n-1} FF^T \tag{14}$$

The eigenvalues and eigenvectors of R were calculated to acquire two eigenvectors corresponding to the largest two eigenvalues. The matrix combined by the two eigenvectors was multiplied by the original normalized matrix:

$$\begin{bmatrix} \tilde{x} \\ \tilde{y} \end{bmatrix} = \begin{bmatrix} \alpha_1 \\ \alpha_2 \end{bmatrix} F \tag{15}$$

In which $\tilde{x}, \tilde{y} \in \mathbb{R}^{1 \times m}$ were the abscissa and ordinate of all dimensionality-reduced data, respectively. The original image and the surface normal map in the dataset were input into the feature extractor separately to get:

$$\vec{f}_i^s, \vec{f}_i^r \in \mathbb{R}^{n \times 1}, i = 1, 2, \ldots, m_p \tag{16}$$

The two vectors were stitched into a matrix:

$$F^a = [\vec{f}_1^s, \vec{f}_2^s, \ldots, \vec{f}_i^s, \vec{f}_1^r, \vec{f}_2^r, \ldots, \vec{f}_i^r] \in \mathbb{R}^{n \times 2m_p}, i = 1, 2, \ldots, m_p \tag{17}$$

Principal component analysis was performed on F^a according to the above method to obtain dimensionality-reduced representations from different domains. Figure 6 showed the results for different kinds of complex features.

It was easy to find from the experimental results that the features from the normal map extracted by the DISS method are more indistinguishable from the features from the original image. On the contrary, the features extracted by RSS and NRSS methods carried obvious domain information. For example, the features from the normal map extracted by the RSS and NRSS methods were clustered together, compared with the features from the original image were more divergent. The features extracted by the DISS method did not have this characteristic. Besides, the features extracted from the normal map by the RSS and NRSS methods had a smaller abscissa, while the features of the original image had a larger abscissa. This pattern was not obvious in the features extracted by the DISS method. Overall, the feature distributions extracted by the DISS method for inputs from the two domains were more similar. In contrast, the features extracted by RSS and NRSS carry obvious domain information. In the training of the pre-task, the amount of original image data was much larger than the amount of normal map data. How to effectively utilize the features extracted from the original image was the key issue of self-supervised learning in this research. The features extracted by the DISS method did not carry obvious domain information, which enabled the features learned from raw images to be used in downstream tasks using normal maps as well. The features extracted by the RSS and NRSS methods had obvious domain information, and the distribution of the extracted features in the two domains had obvious distance, which made the features learned from the original image difficult to be used by downstream tasks.

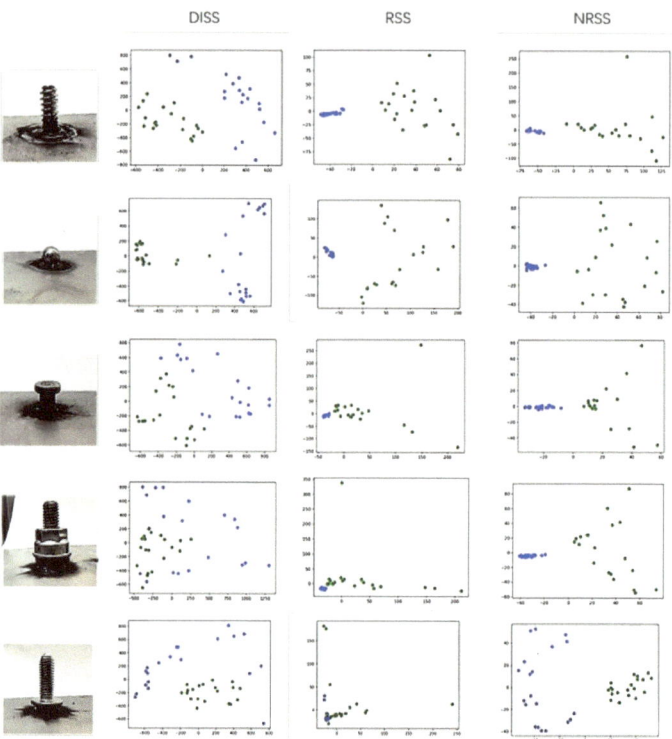

Figure 6. Principal component analysis of extracted features from two inputs.

4. Conclusions

Aiming at the limited training samples of the complex feature detection technology of body-in-white, the few-shot learning technique based on solving the domain gap problem in this paper can effectively reduce the dependence of the complex feature measurement system on the labeled normal map. The proposed self-supervised learning and domain-adaptive learning method had 1.32 pixels lower error than the supervised learning method using normal maps, 0.64 pixels lower than the self-supervised learning method using the original image, and 0.51 pixels lower than the self-supervised learning method using normal maps. In extracting feature principal component analysis, the representations extracted by our method do not exhibit obvious domain information, which proves the effectiveness of our method. The results of this paper are of great significance for improving the data efficiency of complex feature detection methods and reducing implementation costs.

Detection of fusion of complex features and simple features is the direction of further research. There may be simple features with complex features in areas other than automobile body-in-white dimension control. For example, it only has complex geometry, but no complex reflection properties. How to use this characteristic to develop a higher performance and lower cost detection system is a topic worthy of further study.

Author Contributions: Conceptualization, C.L.; methodology, J.L. and J.G.; software, K.S.; validation, C.L.; formal analysis, C.L.; investigation, C.L.; resources, J.L. and J.G.; data curation, K.S.; writ-ing—original draft preparation, C.L.; writing—review and editing, L.Y.; visualization, C.L.; su-pervision, J.G.; project administration, C.L.; funding acquisition, C.L. All authors have read and agreed to the published version of the manuscript.

Funding: This research was funded by State Key Laboratory of Mechanical Behavior and System Safety of Traffic Engineering Structures Independent Project, grant number ZZ2020-37 (Chuang Liu), the Science Research Project of the Education Department of Hebei Province, Grant No. ZD2021093 (Kang Su), National key Research and development Program: 2020YFB1709502 (Jie Li) and National key Research and development Program: 2020YFB1709504 (Jingbo Guo).

Institutional Review Board Statement: Not applicable.

Informed Consent Statement: Not applicable.

Data Availability Statement: Not applicable.

Acknowledgments: The authors thank Northeastern University Machine Vision Laboratory for technical support and help in this research, and the anonymous reviewers and copy editor for valuable comments proposed.

Conflicts of Interest: The authors declare no conflict of interest.

References

1. Liang, J.; Xu, K.; Xu, J. Sub-Pixel Feature Extraction and Edge Detection in 3-D Measuring Using Structured Lights. *Chin. J. Mech. Eng.* **2004**, *12*, 100–103. [CrossRef]
2. Kiraci, E. Moving towards in-line metrology: Evaluation of a Laser Radar system for in-line dimensional inspection for automotive assembly systems. *Int. J. Adv. Manuf. Technol.* **2017**, *91*, 69–78. [CrossRef]
3. Yue, H.; Dantanarayana, H.; Wu, Y. Reduction of systematic errors in structured light metrology at discontinuities in surface reflectivity. *Opt. Lasers Eng.* **2019**, *112*, 68–76. [CrossRef]
4. Carreira, J.; Agrawal, P.; Fragkiadaki, K. Human pose estimation with iterative error feedback. In Proceedings of the IEEE Conference on Computer Vision and Pattern Recognition 2016, Las Vegas, NV, USA, 27–30 June 2016; pp. 4733–4742.
5. Luo, Z.; Zhang, K.; Wang, Z. 3D pose estimation of large and complicated workpieces based on binocular stereo vision. *Appl. Opt.* **2017**, *56*, 6822–6836. [CrossRef] [PubMed]
6. Dang, Q.; Yin, J.; Wang, B. Deep learning based 2d human pose estimation: A survey. *Tsinghua Sci. Technol.* **2019**, *24*, 663–676. [CrossRef]
7. Jing, L.; Tian, Y. Self-supervised visual feature learning with deep neural networks: A survey. *IEEE Trans. Pattern Anal. Mach. Intell.* **2020**, *43*, 4037–4058. [CrossRef] [PubMed]
8. Doersch, C.; Gupta, A.; Efros, A. Unsupervised visual representation learning by context prediction. In Proceedings of the IEEE International Conference on Computer Vision, Santiago, Chile, 13 December 2015.
9. Noroozi, M.; Favaro, P. Unsupervised learning of visual representations by solving jigsaw puzzles. In Proceedings of the European Conference on Computer Vision, Amsterdam, The Netherlands, 8 October 2016.
10. Caron, M.; Bojanowski, P.; Joulin, A. Deep clustering for unsupervised learning of visual features. In Proceedings of the European Conference on Computer Vision, Munich, Germany, 8 September 2018.
11. Pathak, D.; Krahenbuhl, P.; Donahue, J. Context encoders: Feature learning by inpainting. In Proceedings of the IEEE Conference on Computer Vision and Pattern Recognition 2016, Las Vegas, NV, USA, 27–30 June 2016; pp. 2536–2544.
12. Zhang, R.; Isola, P.; Efros, A. Colorful image colorization. In Proceedings of the European Conference on Computer Vision, Amsterdam, The Netherlands, 8 October 2016.
13. Pathak, D.; Girshick, R.; Dollár, P. Learning features by watching objects move. In Proceedings of the IEEE Conference on Computer Vision and Pattern Recognition 2017, Honolulu, HI, USA, 21–26 July 2017; pp. 2701–2710.
14. Liu, H.; Yan, Y.; Song, K. Efficient Optical Measurement of Welding Studs with Normal Maps and Convolutional Neural Network. *IEEE Trans. Instrum. Meas.* **2020**, *70*, 5000614. [CrossRef]
15. Liu, H.; Yan, Y.; Song, K. Optical challenging feature inline measurement system based on photometric stereo and HON feature extractor. In Proceedings of the Optical Micro- and Nanometrology VII, Strasbourg, France, 18 June 2018.
16. Wilson, G.; Cook, D. A survey of unsupervised deep domain adaptation. *ACM Trans. Intell. Syst. Technol.* **2020**, *11*, 1–46. [CrossRef] [PubMed]

Article

A Variable Attention Nested UNet++ Network-Based NDT X-ray Image Defect Segmentation Method

Jiayin Liu and Jae Ho Kim *

Image and A.I. Laboratory, Department of Electronics Engineering, Pusan National University, Busan 46241, Korea; liujiayinpnu@gmail.com
* Correspondence: jhkim@pusan.ac.kr; Tel.: +82-10-4042-2450

Abstract: In this paper, we describe a new method for non-destructive testing (NDT) X-ray image defect segmentation by introducing a variable attention nested UNet++ network. To further enhance the performance of the faint defect extraction and its clear visibility, a pre-processing method based on pyramid model is also added to the proposed method to effectively perform high dynamic range compression and defect enhancement on the 16-bit raw image. To illustrate its effectiveness and efficiency, we applied the proposed algorithm to the X-ray image defect segmentation problem and carried out extensive experiments. The results support that the proposed method outperforms the existing representative techniques in extracting defect for real X-ray images collected directly from industrial lines, which achieves the better performance with 89.24% IoU, and 94.31% Dice.

Keywords: defect segmentation; X-ray image; variable attention; UNet++

Citation: Liu, J.; Kim, J.H. A Variable Attention Nested UNet++ Network-Based NDT X-ray Image Defect Segmentation Method. *Coatings* **2022**, *12*, 634. https://doi.org/10.3390/coatings12050634

Academic Editor: Kechen Song

Received: 3 April 2022
Accepted: 29 April 2022
Published: 5 May 2022

Publisher's Note: MDPI stays neutral with regard to jurisdictional claims in published maps and institutional affiliations.

Copyright: © 2022 by the authors. Licensee MDPI, Basel, Switzerland. This article is an open access article distributed under the terms and conditions of the Creative Commons Attribution (CC BY) license (https://creativecommons.org/licenses/by/4.0/).

1. Introduction

A variety of non-destructive testing (NDT) methods are used in industrial production, to detect the internal defects of objects, among which NDT methods based on X-ray imaging are most widely used, mainly because of their intuitive imaging and high spatial resolution of imaging, especially for the detection of small defects in complex structural products [1–3]. At present, the use of industrial film is still very large. Its disadvantages include slow imaging and being not environmentally friendly, and the process of preliminary preparation, exposure, film processing, interpretation, etc. is very time-consuming. For complex workpieces, it takes several hours of inspection time to complete the whole process, and most importantly, it is difficult to digitize, which brings difficulties to the storage and retrieval of film. With the development of industrialization, a large number of products need full inspection. The previous use of film imaging for random inspection can no longer meet the current pace of inspection, and so the industry of NDT X-ray imaging has an urgent need for digitalization and automation.

Early digital X-ray imaging technology uses an image intensifier (I.I.) as the imaging device, which can output a digital image of up to 12-bits, using the intensifier to output an analog signal and then a digital camera to enhance and digitize the analog signal. Spatial resolution and contrast sensitivity are the most important imaging indicators in X-ray imaging. However, these two indicators of the image intensifier imaging system are not up to the level of film imaging; coupled with the lack of relevant testing standards to support, its application in the field of industrial inspection is limited. In actual industrial applications, large-scale use of image intensifiers for defect detection occurs only in two fields: automotive wheel castings [4,5] and some electronic devices, and small-scale use of the field of welding seams [6] and cylinder inspection [7]. Other areas with strict requirements for detection choose to use film imaging.

The latest X-ray imaging technology uses digital flat panel detectors as imaging devices, which can output 16-bit digital X-ray images with contrast sensitivity indexes that can exceed film imaging. In recent years, the price of flat-panel detector hardware has been reduced to a level generally accepted by industry, and the development of deep learning-based correlation segmentation and identification algorithms has made automated X-ray imaging-based detection possible. However, the diversity of inspected products and their respective complex structures, as well as small and ambiguous defect structures, make automatic defect detection on 16-bit image data extremely complex and challenging.

Defect detection can be roughly divided into three steps: (a) locating and segmenting defects on the X-ray digital image of the inspected product, (b) identifying and quantitatively analyzing the segmented defects, and (c) judging whether the product is qualified according to the relevant analysis data combined with the inspection process requirements. Step (a) is the most difficult one, and it is very important to accurately and completely segment the defects from the image containing the complex structure of the inspected product for later quantitative analysis.

In general, methods for automatic defect segmentation can be divided into two categories: unsupervised and supervised. A variety of unsupervised techniques are used to extract target defects from complex backgrounds [8,9], including matched filtering, morphological processing, defect tracking, etc. One advantage of unsupervised segmentation methods is that no sample annotation is required; however, the practical performance of these methods is not good, especially for small-sized defects with more blurred edges [10]. However, most of the methods in unsupervised learning are based on traditional image processing algorithms, which are not ideal for detection when encountering various complex and variable practical applications, because it is difficult to extract effective features and summarize prediction rules. For example, in the paper [3], the authors used traditional image processing algorithms, mainly median filter and morphological processes for segmentation of defects. The experimental part of the data is relatively small and does not work well on images with slightly complex backgrounds. In [11,12] the authors used the SDD segmentation algorithm to segment ventricles in MRI images and cells or nanoparticles in microscopic images, which have the best performance among the listed single-threshold segmentation methods. However, due to the complicated structure of the workpieces, the grayscale distribution of image background is not uniform. It can be inferred that methods in [11,12] are not very applicable for the industrial X-ray defect segmentation tasks. For supervised methods, the sample image first needs to be manually labeled to mark the defects in it [13], and then the features are separated into background and defects using a trainable classifier. In most cases, the supervised methods perform better than the unsupervised based methods [14,15]. A recent review of deep learning for general object detection can be found in [16], where the authors provide a comprehensive survey of the recent achievements about deep learning for generic object detection.

In recent years, with the rise of deep learning research, several researchers have introduced supervised methods based on deep learning to the task of defect segmentation in X-ray images [17,18]. A very basic fully convolutional network (FCN) for weld defect identification is presented in [19], in which the author illustrates some inspection results on the GDXray database, but comparisons to other deep learning methods and more in-depth tests are lacking. The U-Net algorithm is built on an FCN, consisting of an encoder and a decoder, and the shape of the network resembles a "U" shape, hence the name "U-Net". U-Net is very different from other common segmentation networks in that U-Net uses a completely different feature fusion method: splicing, where U-Net splices features together in the channel dimension to form thicker features [20–22].

Subsequently, the better-performing UNet++ was developed based on U-Net [23]. UNet++ improves segmentation accuracy through a series of nested, dense jump paths that meet the high accuracy requirements for defect detection [24]. The redesigned jump paths make it easier to optimize feature mapping with semantically similar features. Dense jump connections improve segmentation accuracy and improve gradient flow. Deep supervision allows model complexity tuning to balance speed and performance optimization.

UNet++ has been widely used in biological image segmentation, such as retinal vascular segmentation, liver CT image segmentation [25], lung CT image segmentation [26], ultrasound medical image segmentation [27], COVID-19 infection localization [28], heart CT image segmentation [29], etc., and has achieved good results. There are also applications in other scientific and industrial fields, such as the detection of impact craters on the lunar surface [30], road detection [31], etc.

In recent years, there are also some other excellent deep learning network structures for detecting objects in a variety of application scenarios. Vgg16 is a CNN network that simply superimposes convolutional or fully connected layers with weights to 16 layers. As the size of the input image is limited to $224 \times 224 \times 3$, it is difficult to detect smaller defects at the pixel level, and only locates out with boxes for weld defects, with no quantitative output [32]. A spatial attention bilinear convolutional neural network (SA-BCNN) was introduced and tested against other CNN based methods [33]. Faster R-CNN is recognized to have better performance and much research has been done by many scholars. Based on a basic Faster R-CNN system, Feature Pyramid Network (FPN) shows significant improvement as a generic feature extractor in several applications [34]. It achieved state-of-the-art single-model results on the COCO detection benchmark without bells and whistles, surpassing all existing single-model entries including those from the COCO 2016 challenge winners at the time. A method based on Feature Pyramid Network (FPN) and its subsequent improvements was used to detect defects in radiographic images of casting aluminum parts [35,36]. The experimental results outperformed Faster R-CNN; it was an instance segmentation method that could not segment the defects at pixel level and thus could not produce quantitative defect detection results. A very similar pyramid approach is used in networks for deep learning [37], wherein the spatial pyramid pooling is used to remove the uniform limitation on the size of the input image; the pyramid method is a common and useful tool for adapting to different resolutions and scales. Mask R-CNN is introduced through extending Faster R-CNN by adding a branch for predicting an object mask in parallel with the existing branch for bounding box recognition so that it can efficiently detects objects in an image while simultaneously generating a high-quality segmentation mask for each instance, which means it is more suitable for quantitative defect detection [38].

There is no available large-scale defects database of X-ray images, and many scholars use Generative Adversarial Networks (GAN) to generate more defect data for learning in networks such as CNN. A CNN-based method for X-ray prohibited item recognition has been proposed [39], and additionally, generative adversarial networks (GANs) are used for data augmentation. In another CNN-based casting defect detection work [40], the author builds the dataset by using synthetic defects, which are simulated using 3D ellipsoidal models and Generative Adversarial Networks (GAN). This is done not only for X-ray images: in [41] the authors also use GAN to generate more visible defect images to improve defect detection.

In this paper, we propose a novel variable attention-based nested segmentation network that improves the lower segmentation accuracy of the standard UNet++ network using fixed perceptual field convolution. It can automatically adjust the perceptual field of the network by the attention mechanism to more effectively utilize the spatial information extracted at different scales and introduces an attention mechanism between the nested convolutional blocks so that features extracted at different levels can be selected for merging relevant to the segmentation task, as a way to improve the defect segmentation effect of the whole network on sample images with complex background structures. In addition, a new image pre-processing algorithm based on the pyramid model is proposed in this paper, which effectively performs high dynamic range compression and defect enhancement on the original 16-bit images. The grayscale distribution of the processed image is more balanced, and the faint defects that are not easily detected by human eyes are clearly visible, which is convenient for manual labeling.

We further describe the comparison between the proposed method and other related typical methods involved, and to see the advantages and disadvantages of each method more clearly, we summarize them in a table as follows in Table 1.

Table 1. The comparison between the proposed method and other industrial X-ray defect detection methods.

Types	Methods and Brief Description	Advantages	Disadvantages
Unsupervised	Median and morphological filters, Ref [12]	Very easy to implement.	Does not work well on images with slightly complex backgrounds.
	Defect detection based on traditional algorithms, Ref [4]	Better performance than simple morphological and threshold.	Poor results on complex background images.
Supervised	LBP descriptor with an SVM-linear classifier, Ref [16]	Simple features and framework structure.	Not a pixel-level defect segmentation.
	U-Net, Ref [21]	Thicker features, defect fusion at different scales.	The evaluation score of 80% indicates that the system needs modification for better performance.
	Proposed method	Developed HDR pre-processing to enhance more details, large quantity and huge variety of images for training	A little bit time-consuming; it is still acceptable for practical applications.

The experimental results show that after combining the image pre-processing algorithm, the variable attention-based nested UNet++ network proposed in this paper has better detection effect and higher accuracy for X-ray image defects than the other selected network. It was observed that the proposed segmentation method exhibits the top segmentation performance, which holds the leading position with 89.24% IoU, and 94.31% Dice.

2. Methods Section: Algorithm

Using the Pytorch deep learning framework, a variable attention-based nested segmentation network with selective kernel convolution was built. The structure of the proposed network in this paper is based on the U-Net++ network with a nested architecture. For the convolutional module in the network, an improved SK module was used instead of the traditional ordinary convolutional module, and an attention mechanism is introduced between the nested convolutional blocks of the network so that features extracted at different levels can be selectively merged to improve the efficiency of propagating semantic information through jump connections. The proposed network suppresses background regions that are irrelevant to the segmentation task, while having the ability to increase the weight of the target region, which in turn improves the accurate segmentation of defects.

The flow chart of the entire defect detection method is shown in Figure 1, the main important process is divided into: input image, pre-processing, automatic defect detection, and output detection result. The detailed description of each module will be detailed in the subsequent part of this chapter.

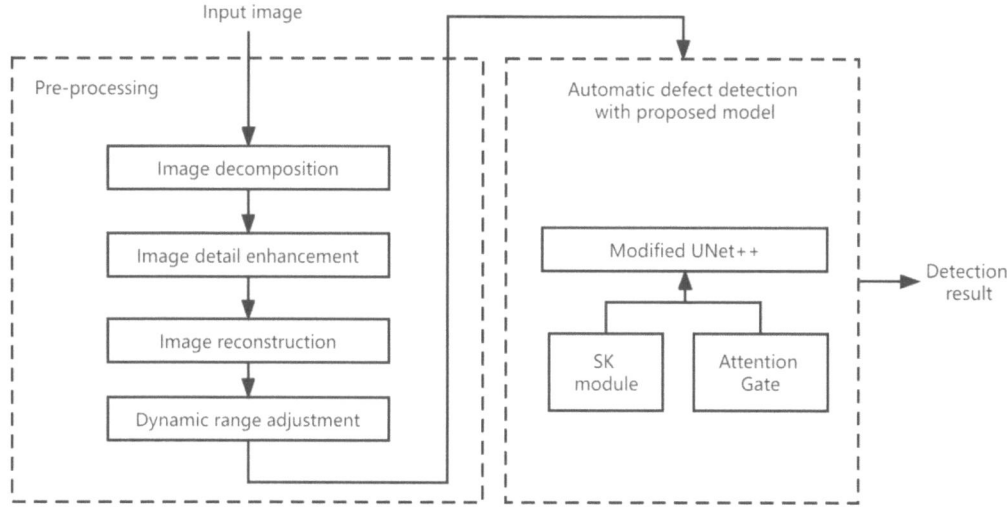

Figure 1. The flow chart of the entire defect detection method.

2.1. Image Pre-Processing

The original image acquired by the X-ray flat-panel detector is a 16-bit grayscale image with a large dynamic range, and the difference between the grayscale values of the defects themselves and those of the background is small, and the grayscale distribution of the background is not uniform due to the inherent X-ray beam-hardening imaging characteristics. We propose a new image pre-processing algorithm based on the pyramid model to effectively perform high dynamic range compression and defect enhancement on the 16-bit raw image. The grayscale distribution of the processed image is more balanced, and the faint defects that are not easily detected by human eyes are clearly visible, which is convenient for manual labeling.

The core idea of our proposed pre-processing algorithm is to decompose the image into pixels representing individual details of the image, then do enhancement on these pixels, and finally perform inverse reconstruction. We choose the Laplace pyramid function for image decomposition, which meets the following two basic conditions: (1) it must include all levels to represent the structure of any size, and (2) it must be continuous without interruption. The effect of pre-processing is shown in Figure 2. On the left is the original image, and on the right is the pre-processed image, where the faint porosity defects in the weld are clearly visible after processing.

2.1.1. Image Decomposition and Reconstruction

The basic idea of Laplace pyramid function decomposition is that first, the original image is low-pass filtered to reduce the closeness of the pixel-to-pixel connection, interval sampling compresses the image data, which means the image sample density is reduced, then interpolation is performed, and finally the resulting image is subtracted from the original image as the first layer in the Laplace pyramid. Repeating the above operations based on this layer of images expands into a pyramid-shaped multi-scale data structure.

Laplace pyramid is built on Gaussian pyramid and consists of a series of L0, L1, L2, L3, L4, etc. As shown in Figure 3, each L is the set of differences between two adjacent Gaussian pyramids, i.e.,

$$L(t) = G(t) - \text{expand}(G(t+1)), \tag{1}$$

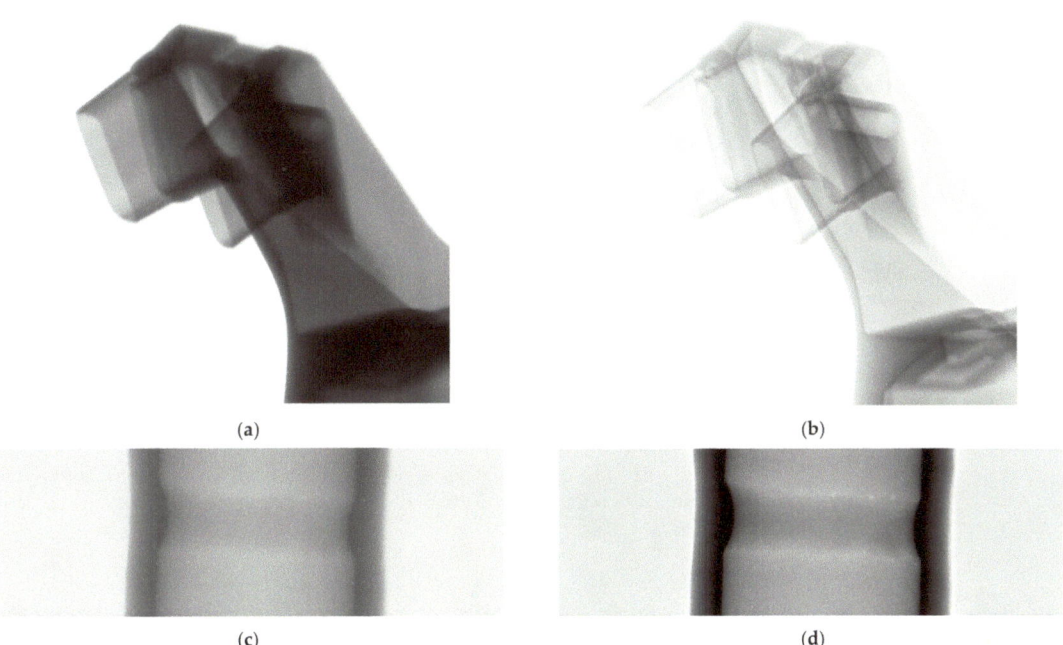

Figure 2. (**a**) Original casting image, (**b**) pre-processed image of (**a**), (**c**) original weld image, (**d**) pre-processed image of (**c**).

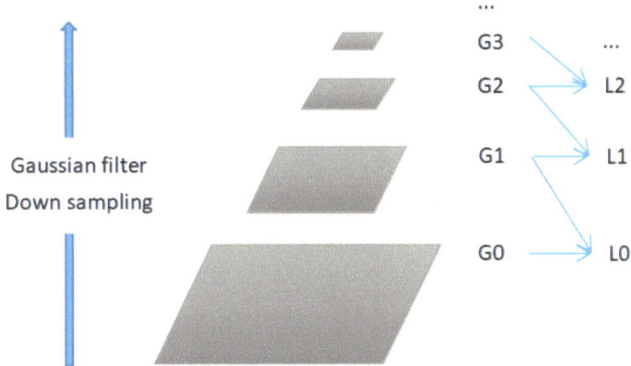

Figure 3. Laplace Pyramid Structure. G0, G1, G2, G3 are the Gaussian pyramid images. In a Gaussian pyramid, subsequent images are weighted down using a Gaussian average and scaled down. L0, L1, L2 are the Laplace pyramid images which save the difference image of the Gaussian images between each levels.

In the image decomposition process, each image is reduced by half (i.e., reduced to one-half of the original sample density) from the previous one, at which point the whole process presents a pyramid data structure. During image reconstruction, the Gaussian pyramid needs to be scaled up and added to the Laplace pyramid of a lower level, i.e.,

$$G(t) = L(t) + \text{expand}(G(t+1)), \tag{2}$$

2.1.2. Image Detail Enhancement

The main purpose is to enhance the details of the image by adjusting the spatial frequency characteristics of the image to highlight the subtle defective features in the image, specifically to achieve

$$Y = X + a \times B(X) \times (X - X1), \quad (3)$$

where Y, X, and X1 represent the pixel values of the resultant image, the original image, and the image after low-pass filtering, (X − X1) represents the high-frequency part of the image, and the coefficient a determines the degree of enhancement of the high-frequency part.

First, the original image is low-pass filtered, that is, smoothed to obtain a smooth image, and then the difference between the original image and the smoothed image to obtain the difference image. The difference image represents the high-frequency information part of the image, usually the edge and detail information part of the image, according to the different data density parts of the image for the corresponding degree of image enhancement, where the coefficient a, function B (X) is given in advance. Here, the coefficient a and function B (X) are given in advance, and will vary according to the image type, image effect requirements, and image data density distribution. The image data density distribution in the algorithm is particularly important, because in the later algorithm Laplace pyramid decomposition, hierarchical enhancement coefficients are related to the density of the image data.

2.1.3. Reduced Dynamic Range

The dynamic range of the image is reduced in the low-frequency part, and the density compensation of the region of interest is implemented:

$$Y = X + a \times (A - X1) \quad (4)$$

where X denotes the original image, X1 denotes the smoothed and filtered image, a is the correction factor where a < 1 (used to control the degree of dynamic compression), and A is a constant.

Firstly, the original image is smoothed by low-pass filtering to obtain a smoothed image, then the difference is made with the given A to obtain a difference image, and then multiplied with the given coefficient a. The meaning is to perform dynamic compression of image data density, i.e., the low frequency part of the image is removed and the high frequency part is retained, because the important detail information of the image often exists in the high frequency. The final image obtained is compared with the original image. The final image is obtained by summing with the original image.

2.2. Defect Segmentation Network

The Pytorch deep learning framework is used to build this network, and the structure of the proposed convolutional neural network in this paper is shown in Figure 4, which is based on the UNet++ network and uses a nested architecture to integrate U-Net of different depths.

The network designed in this paper nests four layers of U-Net as the basic network framework, where encoders and decoders are symmetrically distributed on both sides of the network. All layers of U-Net share one feature extractor so that we only need to train one encoder. A modified SK block is used to replace the traditional convolutional block in the network, and a convolution with a perceptual field of 5 is generated by using two 3×3 convolutions in series in the SK block, which both improves the depth of the network and reduces the computation and number of parameters, as shown in Figure 5. By using the SK block, the perceptual field can be automatically adjusted to make more efficient use of the feature information extracted at different scales. The encoder has a total of five layers, each of which consists of two modified SK blocks + Relu. Each layer undergoes a maximum pooling of size 2×2 with a step size of 2 after feature extraction. Each subsequent layer of the structure is down-sampled in the same order.

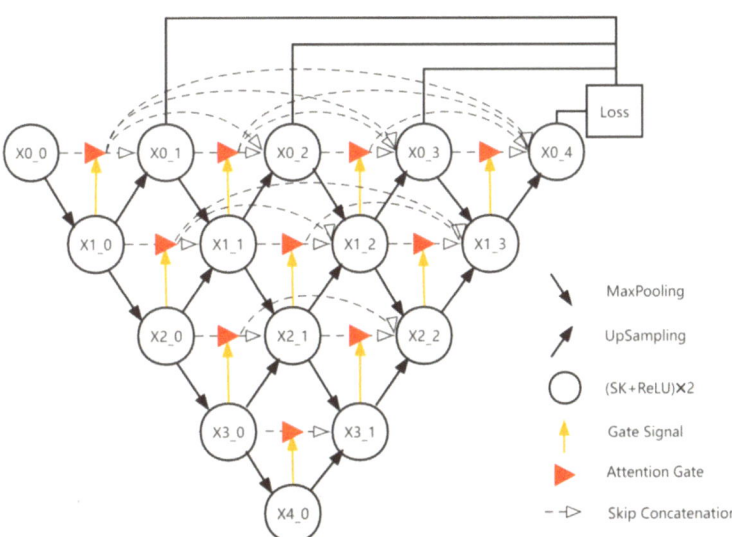

Figure 4. Schematic diagram of the structure of the proposed defect segmentation network. SK block is adopted instead of convolutional block with an attention gate.

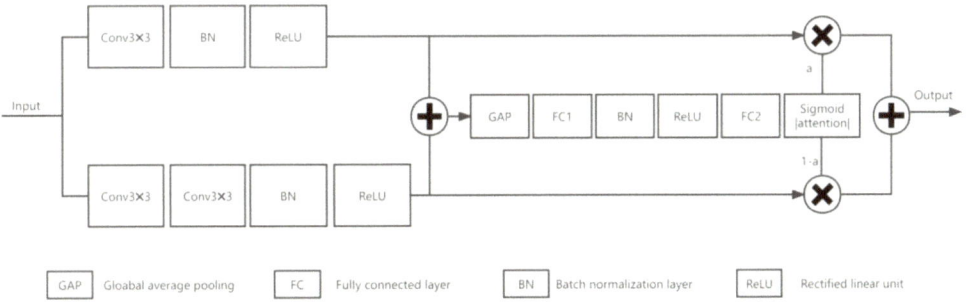

Figure 5. Schematic diagram of the structure of the improved SK module. The SK block consists of two branches. The first one utilizes one conventional 3 × 3 convolutions, while the second one uses two 3 × 3 convolutions to generate a perceptual field of 5.

To focus on features related to the target or goals, we add a simple but effective attention gate to the nested architecture, as shown in Figure 6. This attention gate has two inputs: an up-sampled feature Fg in the decoder and a feature Fx of equal depth in the encoder. The selected signal Fg in the attention gate selects the more useful features from the encoded feature Fx and sends them to the upper decoder.

The contextual information extracted by the encoder is propagated to the decoder of the corresponding layer through a dense jump connection, thus allowing the extraction of more efficient layered features. In the case of dense jump connection, the input of each convolutional block in the decoder consists of two equal-scale feature maps: (1) the intermediate feature map from the output of the previous potential gate along the same depth of the jump connection; and (2) the final feature map from the output of a deeper deconvolution block operation. After receiving and concatenating all the to-be-feature maps, the decoder recovers the image in a bottom-up manner.

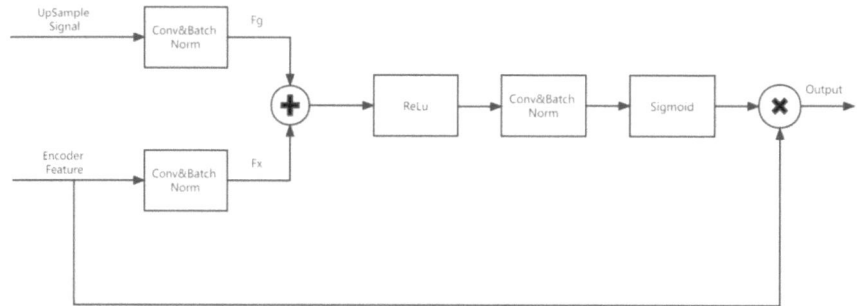

Figure 6. Schematic diagram of the structure of the attention gate. Encoder features are scaled with attention coefficients calculated in the attention gate.

We define the feature mapping to represent the output of the convolutional block, where i represents the depth of the feature in the network and j represents the sequence of convolutional blocks in layer i connected along the jump, as follows.

$$X_{i,j} = \begin{cases} \phi[X_{i-1,j}] & j = 0 \\ \phi\left[\sum_{k=0}^{j-1} Ag(X_{i,k}), UP(X_{i+1,j-1})\right] & j > 0' \end{cases} \quad (5)$$

where $\phi[]$ denotes the concatenated merging of convolutional blocks. $UP[]$ and $Ag[]$ denote up-sampling and attention gate selection, respectively.

Deep supervision is introduced in the network structure by attaching a 1×1 convolution with a C kernel and a sigmoid activation function to the outputs of nodes X0_1, X0_2, X0_3 and X0_4, where C is the number of classes in which a given dataset is available. A hybrid segmentation loss is then defined for each semantic scale, including pixel-level cross-entropy and Dice coefficient loss. The hybrid loss can take advantage of the two loss functions: smooth gradients and class imbalance handling. It is defined as follows.

$$L(Y,P) = -\frac{1}{N} \sum_{c=1}^{C} \sum_{n=1}^{N} (y_{n,c} \log p_{n,c} + \frac{2y_{n,c}p_{n,c}}{y_{n,c}^2 + p_{n,c}^2}), \quad (6)$$

of which $y_{n,c} \in Y$ and $p_{n,c} \in P$ denote the target labels and predicted probabilities for class c and n pixels in a batch, and N denotes the number of pixels in a batch.

Because the output of each sub-network is already the segmentation result of the image during deep supervision, we can cut out those redundant parts if the output result of the smaller sub-networks is already good enough.

The main advantages of the defect segmentation method proposed in this paper include the following:

1. The traditional convolutional blocks are replaced by SK blocks, and the convolutional blocks with a perceptual field of 5 in the SK blocks are replaced by two 3x3 convolutions in series, which not only improve the depth of the network but also reduce the computation and the number of parameters. By using the SK block, the perceptual field can be automatically adjusted to utilize the feature information extracted at different scales more effectively.
2. Adding attention gates between nested convolutional blocks enables increase of the weight of the target region while suppressing background regions that are not relevant to the segmentation task.
3. It enables model pruning during testing by introducing deep supervision, which can reduce a large number of model parameters and thus speed up the model segmentation.

3. Experimental Section

In this section, both quantitative and qualitative results are reported with an extensive set of comparative evaluations for defect segmentation.

3.1. Experimental Equipment and Database

Unlike the vast database of publicly available medical X-ray images, currently, the only publicly accessible X-ray imaging database for industrial NDT is GDXray, which was published by D. Mery in 2015 [42]. The database includes five groups of X-ray images: castings, welds, baggage, natural objects, and settings. However, the image number of GDXray's database is too small, with only 2727 images from different angles on 67 samples of castings, and only 88 images from different positions on three samples of welds. In addition, most of the images were taken with the Image Intensifier (I.I.) and saved in 8-bit BMP or JPG format. Thus, the image quality and spatial resolution were not high enough, and could not be adapted to the contemporary needs of industrial inspection. Figure 7 shows almost 40 images generated at different shooting angles from one single aluminum casting wheel in GDXray's database. Therefore, we needed to build our own database.

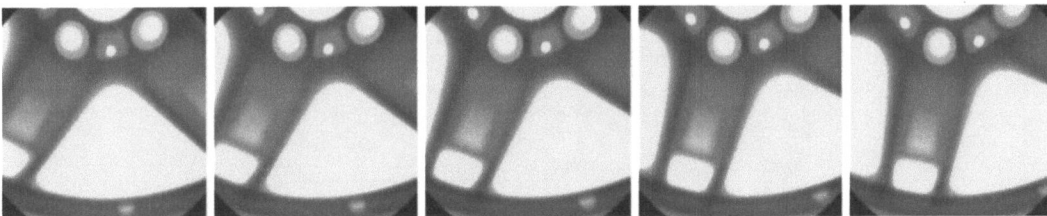

Figure 7. Selected images from GDXray's database [42].

We have cooperated with Deepsea Precision Co. Ltd. (Shenzhen, China) [43], a manufacturer specializing in X-ray inspection equipment, to collect a large amount of high-quality, high-resolution X-ray image data from actual production lines and laboratories over a period of three years. As shown in Figure 8, this is industrial X-ray inspection equipment from Deepsea Precision Co. Ltd., and our database was collected from this equipment and more than a dozen other similar machines. The basic configuration of the core imaging chain components from this typical X-ray inspection equipment of Figure 8a is shown in Table 2.

Table 2. The basic configuration of the core imaging chain components from the X-ray inspection equipment of Figure 8a.

Device Name	Brand/Model	Basic Configuration
X-ray emission device (macro-focus)	Gulmay/CF500	500 kV, focus size 0.4/1.0 mm
X-ray emission device (micro-focus)	WorX/XWT-225-CT	225 kV, focal size 5 μm
X-ray receiver device (flat panel detector)	Deepsea/DS4343HR	430 mm * 430 mm, pixel size 139 μm
Workstation Software	Deepsea/DeepVISION	GPU-based architecture

The database we created is named DSXImage, and its architecture is shown in Table 3. It is divided into two main types, castings and welds, and the image format is fully compliant with the industrial NDT standard ASTM E2339-15 [44] and is saved in 16-bit DICONDE format. The breakdown within each main category is as follows.

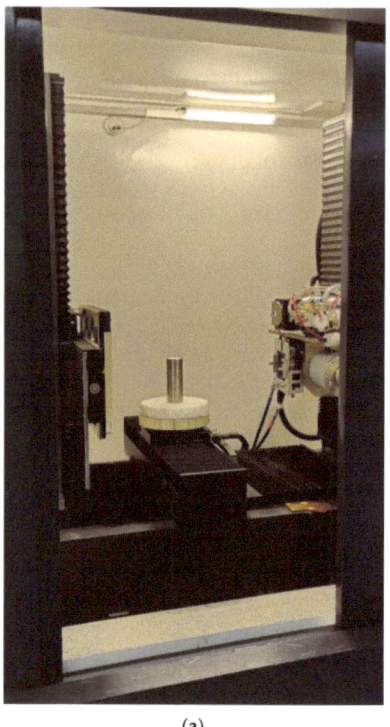
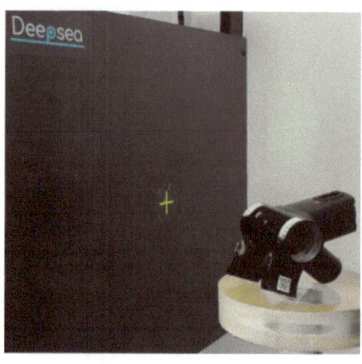
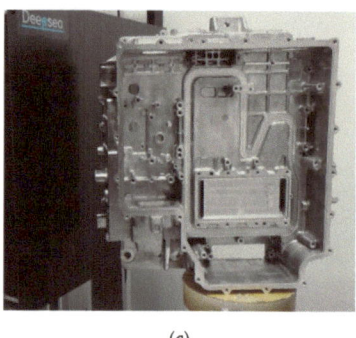

(a)

(b)

(c)

Figure 8. Some typical X-ray inspection equipment from Deepsea. (**a**) A 500 kV X-ray inspection machine, steel weld sample; (**b**) a 225 kV X-ray inspection machine, aluminum casting sample from a bicycle; (**c**) a 225 kV X-ray inspection machine, aluminum casting sample from automobile.

Table 3. DSXImage database architecture.

Types	Sample Name	Amount
Castings	engine cylinder heads	10,500
	steering knuckles	8602
	shock absorbers	15,610
	valves	9500
	power supply housings	6500
	3C products and others	18,805
Welds	rocket engine ducts	12,655
	aircraft seat bases	5431
	gas pipe welds	25,820
	steel welds from 3C and others	12,907

1. Most of the castings are aluminum alloy parts for automobiles, such as engine cylinder heads, steering knuckles, shock absorbers, valves, power supply housings, etc. A small portion are magnesium alloy parts for 3C products, such as Bluetooth headset metal frames, laptop bezels, etc.
2. Welds are stainless steel/titanium alloy, collected from rocket engine ducts, aircraft seat bases, gas pipe welds, etc. A small portion are steel welds from 3C.

The DSXImage database is still in continuous improvement; however, the existing size of the database is sufficient for training and evaluation of deep learning networks. The database may be released to the public at the right time.

The project team worked with Deepsea to generate the DSXImage training dataset using the open-source labeling tool Labelme., which is a graphical image annotation tool inspired by MIT. The user can use it to easily implement image annotation work for vision tasks such as classification, detection, and segmentation, etc. The annotation result of one sample consists of three images and two label files. The architecture is summarized in Table 4. The image annotation work is a time-consuming and labor-intensive task that took nearly a year to label the entire database. The specific annotation process and output are summarized roughly as follows.

Table 4. The architecture of one annotation result, basic configuration of the core imaging chain components from the X-ray inspection equipment of Figure 7a.

The Original Image	The Label Image	The Label Visualization Image

1. Open the image, perform manual annotation, outline each defect and save it as a json label file.
2. Open a json file and convert the json into a mask label image.
3. After the conversion is completed, a label folder is generated, including the original image img.png, label image label.png, label visualization image label_viz.png, txt file of label name, and yaml format label name file.

The design principle of the database is to conform to the current mainstream image specifications and try to take into account the previous image specifications so that not only can previous algorithms use the database, but future algorithms can also be adapted by simple expansion and upgrade of the current database.

1. Image bits. Most modern flat panel detectors output 16-bit images, older detectors are 12 or 14 bit, and earlier image intensifiers are 8 or 10 bit. The database uses the original 16-bit image with the addition of a pre-processed 16-bit image and an 8-bit image.
2. Image resolution. Depending on the specifications of the flat panel detector, there is no uniform specification for the resolution of the images in the database. The pixel sizes of most images are 3072 × 3072, 1536 × 1536, 2000 × 2000, which users can choose according to their needs and can even be freely cropped if necessary.
3. Image format. The 10-bit, 12-bit, 14-bit, and 16-bit images in the database use the industry standard image format, i.e., DICONDE format, and also keep a copy in TIFF format. 8-bit images are saved in PNG format.

3.2. Experimental Setup

The defect segmentation task was conducted over the DSXImage dataset comprising 126,330 sample X-ray images with corresponding image annotation. In all our experiments, we assumed an 80/20 split for train and test purposes respectively. Besides, 20% of training data was used as a validation set for model selection and to avoid overfitting. PyTorch library with Python 3.7 was used to train and evaluate the variable attention nested UNet++ network, running on a PC with Intel® Core™ i9-9900KF CPU at 3.6 GHz (Santa Clara, CA, USA), with 64 GB RAM, and with a 12 GB NVIDIA GeForce GTX 2060 GPU card (Santa Clara, CA, USA) [45,46].

Some of the original images in the DSXImage datasets are shown in Figure 9.

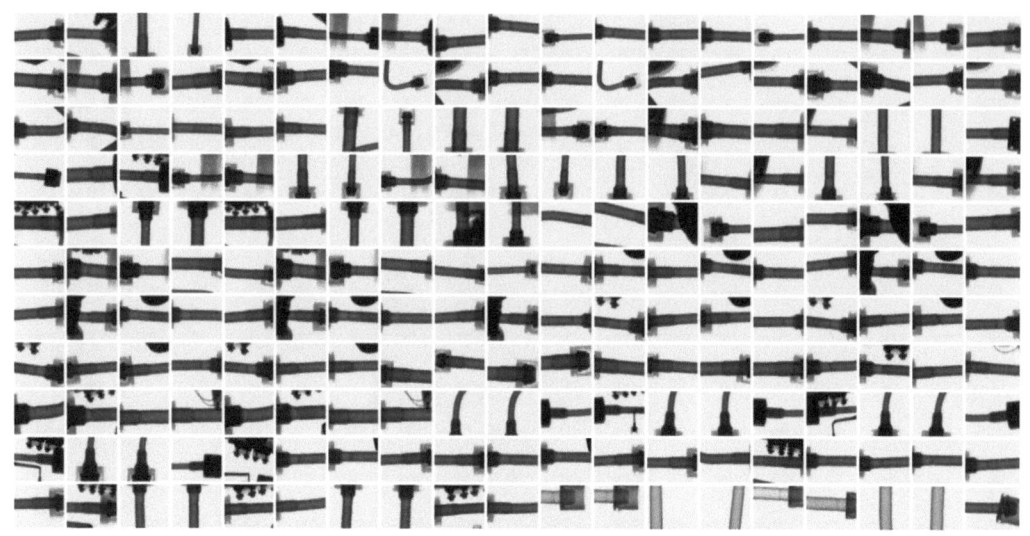

Figure 9. Some of the original images in the DSXImage dataset.

3.3. Segmentation Evaluation Metrics

The performance of the defect segmentation networks was evaluated using two evaluation metrics, namely, Dice (Dice Similarity Coefficient) and IoU (Intersection over Union) [28]. The corresponding equations are as follows.

$$Dice = 2 * TP/(2 * TP + FP + FN), \qquad (7)$$

$$IoU = TP/(TP + FP + FN), \qquad (8)$$

Here, TP, TN, FP, FN represent the true positive, true negative, false positive, and false negative, respectively. Both IoU and Dice are statistical measures of spatial overlap between the binary ground-truth and the predicted segmentation masks, where the main difference is that Dice considers double weight for TP pixels (true defect predictions) compared to IoU.

3.4. Segmentation Results

The performance of the all the selected segmentation methods over the test set is tabulated in Table 5. We have selected some classical and recent deep-learning-based segmentation algorithms with better performance. For all models, it was observed that the proposed segmentation method exhibits the top segmentation performance, holding the leading position with 89.24% IoU, and 94.31% Dice. Some examples of segmentation results obtained by different methods listed in Table 5 are shown in Figures 10 and 11. Figure 12 shows more examples of the segmentation results of the proposed method. Furthermore, the proposed method has been customized for different applications as illustrated in Figure 13.

Table 5. The performance of the all the selected segmentation method over the test set—DSXImage database.

Models	IoU	Dice	#Parameters	Time per Epoch
Median and morphological filters, Ref [3]	0.3012	0.4630	–	–
Defect detection based on traditional algorithms, Ref [4]	0.4155	0.5871	–	–
Mask R-CNN	0.7360	0.8479	25.3 M	38 s
U-Net	0.7251	0.8406	32.5 M	16 s
Standard UNet++	0.7629	0.8656	35.6 M	39 s
Proposed Method	0.8924	0.9431	36.8 M	42 s

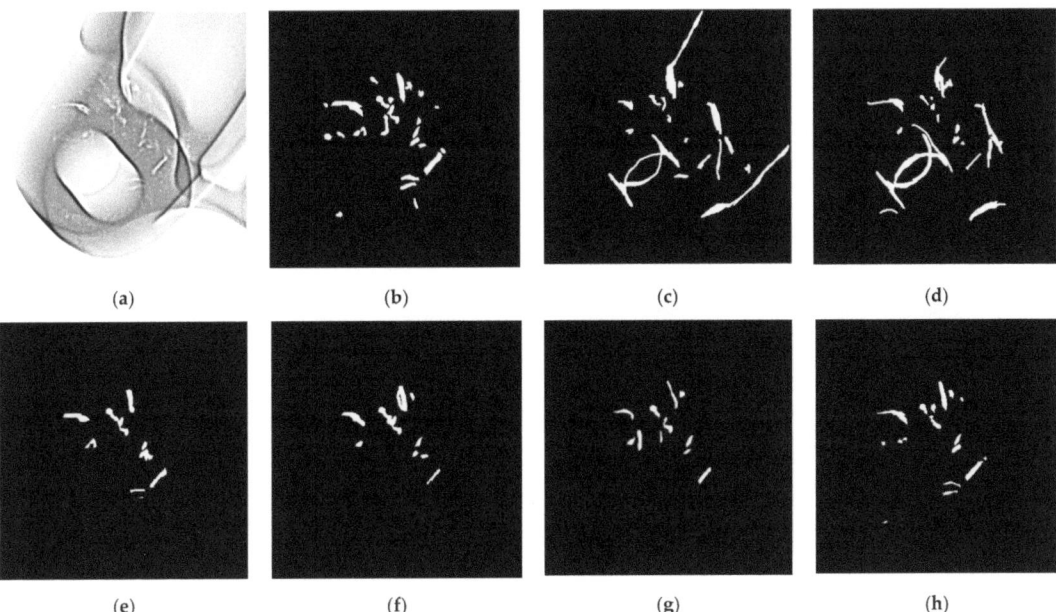

Figure 10. A typical example of defect segmentation: (**a**) the original image, (**b**) ground-truth image, (**c**) result of median and morphological filters, Reference [3], (**d**) result of a traditional defect detection algorithm, Reference [4], (**e**) result of Mask R-CNN, (**f**) result of UNet, (**g**) result of standard UNet++, (**h**) result of proposed method.

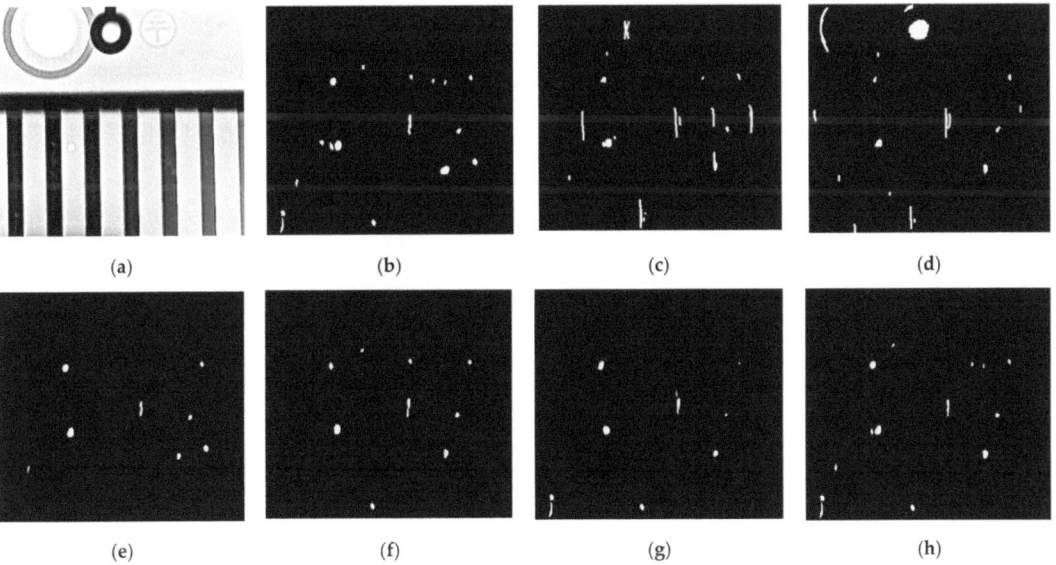

Figure 11. Another example of defect segmentation: (**a**) the original image, (**b**) ground-truth image, (**c**) result of median and morphological filters, Reference [3], (**d**) result of a traditional defect detection algorithm, Reference [4], (**e**) result of Mask R-CNN, (**f**) result of UNet, (**g**) result of standard UNet++, (**h**) result of proposed method.

(a) (b) (c) (d)

(e) (f) (g) (h)

(i) (j) (k) (l)

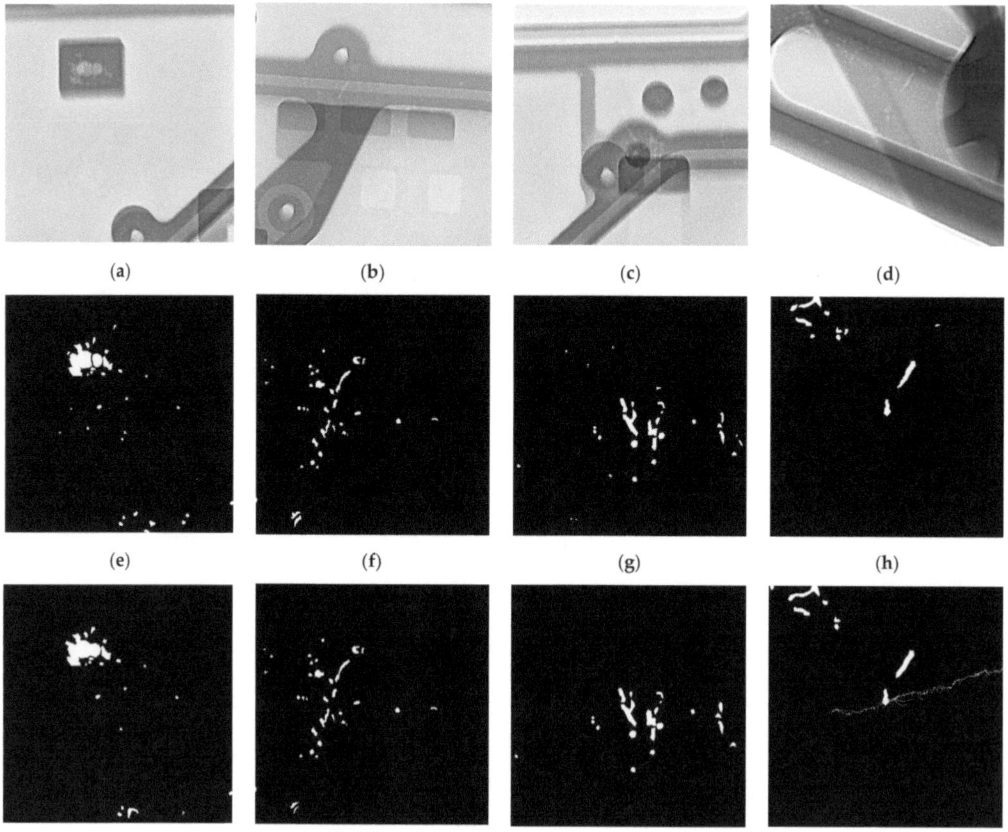

Figure 12. Segmentation examples of proposed method. (**a**–**d**) are the original images; (**e**–**h**) are the ground-truth images; (**i**–**l**) are the corresponding segmentation results.

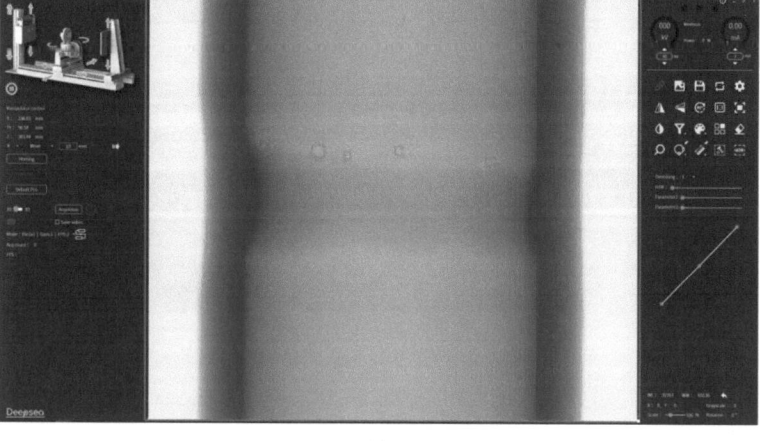

(**a**)

Figure 13. *Cont.*

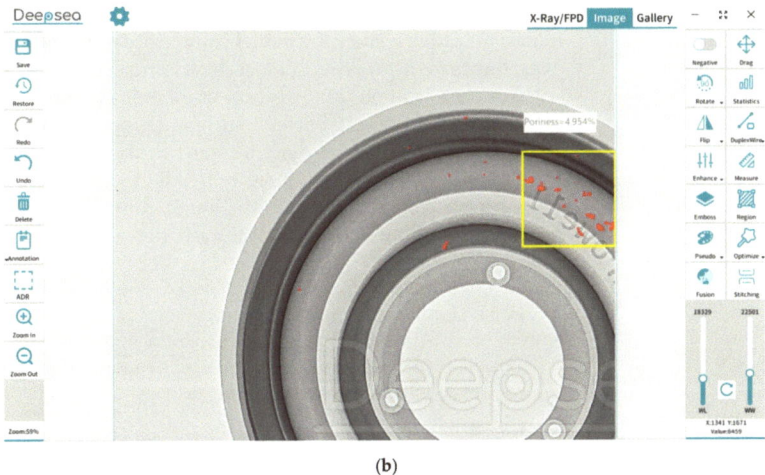

(b)

Figure 13. Applications of proposed method: (**a**) automatic welding defect detection, (**b**) automatic casting defect detection.

4. Discussion

In this section, we will discuss the results obtained by the proposed defect detection system as well as other methods including some deep learning models and conventional defect detection methods. The segmentation results accordingly are shown in Figures 10 and 11. According to our experimental results, deep learning methods provide better segmentation performance in comparison to conventional image processing methods. In general, conventional methods generate high false positive rate due to the characteristics of industrial X-ray images.

Figure 12 shows that our model performs well on complex images and is able to detect different kind of defects. The performance of the proposed model and some other models are evaluated in Table 5. Mask R-CNN and U-Net showed similar accuracy in the experiments. In addition, the proposed modified UNet++ model outperformed the standard UNet++. Due to the improvement of image quality through proposed pre-processing, our defect segmentation method achieves 89.24% and 94.31% in terms of IoU and Dice, respectively.

Although the overall performance of the proposed method has been demonstrated, there are still some missed detections in the experiments. This is mainly because some thin defects are slightly stronger than the background, increasing the difficulty of detection. Moreover, some incorrect detections were also observed due to the similarity between the edges or holes of the parts and the defects.

5. Conclusions

In this paper, we describe a new method for X-ray image defect segmentation by introducing a variable attention nested UNet++ network. In comparison with the existing techniques, the proposed algorithm has the following features: (i) a pre-processing method based on pyramid model is proposed to further enhance the performance of the faint defect extraction and its clear visibility; (ii) the traditional convolutional blocks are replaced by SK blocks, and adding attention gates between nested convolutional blocks enables increase in the weight of the target region while suppressing background regions that are not relevant to the segmentation task; (iii) it enables model pruning during testing by introducing deep supervision, which can reduce a large number of model parameters and thus speed up the model segmentation. Moreover, the proposed segmentation method proved reliable in localizing defects from the DSXImage database, achieving IoU and Dice values of 89.24% and 94.31%, respectively.

In the future, we plan to explore robust quantization and model compression techniques to further reduce the model complexity and accelerate the inference process. Moreover, the current proposed defect detection system is developed to identify the defects without classification. In the future work, this system will be modified and improved to identify different types of defects such as gas holes, gas porosity, and shrinkage cavity.

We will continue to push forward in the construction of the database, and with the consent of Deepsea, we hope to make the database public in the near future so that more people can participate in the work, continue to improve the defect segmentation performance, and make the algorithms better serve the industry.

Author Contributions: Conceptualization, J.L. and J.H.K.; methodology, J.L.; software, J.L.; validation, J.L.; formal analysis, J.L.; investigation, J.L. and J.H.K.; resources, J.L.; data curation, J.L.; writing—original draft preparation, J.L.; writing—review and editing, J.H.K.; visualization, J.L.; supervision, J.H.K.; project administration, J.L.; funding acquisition, J.L. All authors have read and agreed to the published version of the manuscript.

Funding: This research was supported by Guangdong Science and Technology Project, grant number 2020A0505100012, 2017B020210005 and BK21FOUR, Creative Human Resource Education and Research Programs for ICT Convergence in the 4th Industrial Revolution.

Acknowledgments: Thanks to Deepsea Precision Co. Ltd. and for their X-ray inspection equipment and support for the large number of samples tested over two years. Thanks to all the people who contributed to the database annotation. Special thanks to Stan, a colleague of J.L., who did a lot of work in organizing the data.

Conflicts of Interest: The authors declare no conflict of interest.

References

1. Mery, D. Automatic Defect Detection in Aluminium Castings using Radioscopic Image Sequences. *Met. Cast. Technol.* **2003**, *49*, 1.
2. Hangai, Y.; Kawato, D.; Ando, M.; Ohashi, M.; Morisada, Y.; Ogura, T.; Fujii, H.; Nagahiro, R.; Amagai, K.; Utsunomiya, T.; et al. Nondestructive observation of pores during press forming of aluminum foam by X-ray radiography. *Mater. Charact.* **2022**, *170*, 110631. [CrossRef]
3. Fadel, N.; Al-Hameed, W.; Ahmed, M.K. Non Destructive Testing For Detection abnormal Object in The X-Ray Images. *J. Phys. Conf. Ser.* **2020**, *1660*, 012104. [CrossRef]
4. Mery, D.; Filbert, D. Automated flaw detection in aluminum castings based on the tracking of potential defects in a radioscopic image sequence. *IEEE Trans. Robot. Autom.* **2002**, *18*, 890–901. [CrossRef]
5. Mery, D. Automated radioscopic testing of aluminum die castings. *Mater. Eval.* **2006**, *64*, 135–143.
6. Da Silva, R.R.; Mery, D. State-of-the-art of weld seam inspection by radiographic testing. *Image Processing Mater. Eval.* **2007**, *6*, 643–647.
7. Connolly, C. X-ray systems for security and industrial inspection. *Sens. Rev.* **2008**, *28*, 194–198. [CrossRef]
8. Pieringer, C.; Mery, D. Flaw detection in aluminium die castings using simultaneous combination of multiple views. *Insight-Non-Destr. Test. Cond. Monit.* **2010**, *52*, 548–552. [CrossRef]
9. Mery, D.; Lillo, I.; Loebel, H.; Riffo, V.; Soto, A.; Cipriano, A.; Aguilera, J.M. Automated fish bone detection using X-ray imaging. *J. Food Eng.* **2011**, *105*, 485–492. [CrossRef]
10. Mery, D. Inspection of complex objects using multiple-X-ray views. *IEEE/ASME Trans. Mechatron.* **2014**, *20*, 338–347. [CrossRef]
11. Wang, Z. Automatic localization and segmentation of the ventricles in magnetic resonance images. *IEEE Trans. Circuits Syst. Video Technol.* **2020**, *31*, 621–631. [CrossRef]
12. Wang, Z. A new approach for segmentation and quantification of cells or nanoparticles. *IEEE Trans. Ind. Inform.* **2016**, *12*, 962–971. [CrossRef]
13. Mery, D.; Hahn, D.; Hitschfeld, N. Simulation of defects in aluminium castings using CAD models of flaws and real X-ray images. *Insight-Non-Destr. Test. Cond. Monit.* **2005**, *47*, 618–624. [CrossRef]
14. Mery, D.; Arteta, C. Automatic defect recognition in x-ray testing using computer vision. In Proceedings of the 2017 IEEE Winter Conference on Applications of Computer Vision (WACV), Santa Rosa, CA, USA, 24–31 March 2017.
15. Mery, D.; Pieringer, C. *Computer Vision for X-ray Testing: Imaging, Systems, Image Databases, and Algorithms*, 2nd ed; Springer International Publishing: Cham, Switzerland, 2021.
16. Liu, L.; Ouyang, W.; Wang, X.; Fieguth, P.; Chen, J.; Liu, X.; Pietikäinen, M. Deep learning for generic object detection: A survey. *Int. J. Comput. Vis.* **2020**, *128*, 261–318. [CrossRef]
17. Mery, D.; Pieringer, C. Deep Learning in X-ray Testing. In *Computer Vision for X-ray Testing*; Springer: Cham, Switzerland, 2021; pp. 275–336.

18. Mery, D. Aluminum casting inspection using deep object detection methods and simulated ellipsoidal Defects. *Mach. Vis. Appl.* **2021**, *32*, 1–16. [CrossRef]
19. Tokime, R.; Maldague, X.; Perron, L. Automatic Defect Detection for X-Ray inspection: Identifying defects with deep convolutional network. In Proceedings of the Canadian Institute for Non-destructive Evaluation (CINDE), Edmonton, AB, Canada, 18–20 June 2019.
20. Ronneberger, O.; Fischer, P.; Brox, T. U-net: Convolutional networks for Biomedical Image Segmentation. In *International Conference on Medical Image Computing and Computer-Assisted Intervention*; Springer: Cham, Switzerland, 2015.
21. Li, X.; Chen, H.; Qi, X.; Dou, Q.; Fu, C.W.; Heng, P.A. H-DenseUNet: Hybrid densely connected UNet for liver and tumor segmentation from CT volumes. *IEEE Trans. Med. Imaging* **2018**, *37*, 2663–2674. [CrossRef]
22. Tokime, R.B.; Maldague, X.; Perron, L. Automatic Defect Detection for X-ray inspection: A U-Net approach for defect segmentation. In Proceedings of the Digital Imaging and Ultrasonics for NDT Conference 2019, New Orleans, LA, USA, 23–25 July 2019.
23. Zhou, Z.; Rahman Siddiquee, M.M.; Tajbakhsh, N.; Liang, J. Unet++: A nested u-net architecture for medical image segmentation. In *Deep Learning in Medical Image Analysis and Multimodal Learning for Clinical Decision Support*; Springer: Cham, Switzerland, 2018; pp. 3–11.
24. Zhou, Z.; Siddiquee, M.M.R.; Tajbakhsh, N.; Liang, J. Unet++: Redesigning skip connections to exploit multiscale features in image segmentation. *IEEE Trans. Med. Imaging* **2019**, *39*, 1856–1867. [CrossRef]
25. Li, C.; Tan, Y.; Chen, W.; Luo, X.; Gao, Y.; Jia, X.; Wang, Z. Attention unet++: A nested attention-aware u-net for liver ct image segmentation. In Proceedings of the 2020 IEEE International Conference on Image Processing (ICIP), Abu Dhabi, United Arab Emirates, 25–28 October 2020.
26. Gite, S.; Mishra, A.; Kotecha, K. Enhanced lung image segmentation using deep learning. *Neural Comput. Appl.* **2022**, 1–15. [CrossRef]
27. Zhou, Q.; Wang, Q.; Bao, Y.; Kong, L.; Jin, X.; Ou, W. LAEDNet: A Lightweight Attention Encoder–Decoder Network for ultrasound medical image segmentation. *Comput. Electr. Eng.* **2022**, *99*, 107777. [CrossRef]
28. Tahira, A.M.; Chowdhury, M.E.H.; Khandakar, A.; Rahman, T.; Qiblawey, Y.; Khurshid, U.; Kiranyaz, S.; Ibtehaz, N.; Sohel Rahman, M.; Al-Maadeed, S.; et al. COVID-19 infection localization and severity grading from chest X-ray images. *Comput. Biol. Med.* **2021**, *139*, 105002. [CrossRef]
29. Diniz, J.O.B.; Dias Júnior, D.A.; da Cruz, L.B.; da Silva, G.L.F.; Ferreira, J.L. Heart segmentation in planning CT using 2.5 D U-Net++ with attention gate. *Comput. Methods Biomech. Biomed. Eng. Imaging Vis.* **2022**, 1–9, in press. [CrossRef]
30. Jia, Y.; Liu, L.; Zhang, C. Moon Impact Crater Detection Using Nested Attention Mechanism Based UNet++. *IEEE Access* **2021**, *9*, 44107–44116. [CrossRef]
31. Yang, X.; Li, X.; Ye, Y.; Zhang, X.; Zhang, H.; Huang, X.; Zhang, B. Road detection via deep residual dense u-net. In Proceedings of the 2019 International Joint Conference on Neural Networks (IJCNN), Budapest, Hungary, 14–19 July 2019.
32. Liu, B.; Zhang, X.; Gao, Z.; Chen, L. Weld defect images classification with vgg16-based neural network. In *International Forum on Digital TV and Wireless Multimedia Communications*; Springer: Singapore, 2017.
33. Tang, Z.; Tian, E.; Wang, Y.; Wang, L.; Yang, T. Nondestructive defect detection in castings by using spatial attention bilinear convolutional neural network. *IEEE Trans. Ind. Inform.* **2020**, *17*, 82–89. [CrossRef]
34. Lin, T.; Dollár, P.; Girshick, R.; He, K. Feature pyramid networks for object detection. In Proceedings of the IEEE Conference on Computer Vision and Pattern Recognition, Honolulu, HI, USA, 21–26 July 2017.
35. Du, W.; Shen, H.; Fu, J.; Zhang, G.; He, Q. Approaches for improvement of the X-ray image defect detection of automobile casting aluminum parts based on deep learning. *NDT E Int.* **2019**, *107*, 102144. [CrossRef]
36. Du, W.; Shen, H.; Fu, J.; Zhang, G.; Shi, X.; He, Q. Automated detection of defects with low semantic information in X-ray images based on deep learning. *J. Intell. Manuf.* **2021**, *32*, 141–156. [CrossRef]
37. He, K.; Zhang, X.; Ren, S.; Sun, J. Spatial pyramid pooling in deep convolutional networks for visual recognition. *IEEE Trans. Pattern Anal. Mach. Intell.* **2015**, *37*, 1904–1916. [CrossRef]
38. He, K.; Gkioxari, G.; Dollar, P.; Girshick, R. Mask R-CNN. In Proceedings of the IEEE International Conference on Computer Vision, Venice, Italy, 22–29 October 2017.
39. Yang, J.; Zhao, Z.; Zhang, H.; Shi, Y. Data augmentation for X-ray prohibited item images using generative adversarial networks. *IEEE Access* **2019**, *7*, 28894–28902. [CrossRef]
40. Mery, D. Aluminum casting inspection using deep learning: A method based on convolutional neural networks. *J. Nondestruct. Eval.* **2020**, *39*, 1–12. [CrossRef]
41. Niu, S.; Li, B.; Wang, X.; Lin, H. Defect image sample generation with GAN for improving defect recognition. *IEEE Trans. Autom. Sci. Eng.* **2020**, *17*, 1611–1622. [CrossRef]
42. Mery, D.; Riffo, V.; Zscherpel, U.; Mondragón, G.; Lillo, I.; Zuccar, I.; Lobel, H.; Carrasco, M. GDXray: The database of X-ray images for nondestructive testing. *J. Nondestruct. Eval.* **2015**, *34*, 1–12. [CrossRef]
43. Deepsea Precision Co., Ltd. Shenzhen City, China. Available online: http://www.shenhaijingmi.com/ (accessed on 2 April 2022).
44. ASTM E2339-15. *Standard Practice for Digital Imaging and Communication in Nondestructive Evaluation (DICONDE), 2015 Edition.* 1 December 2015. Available online: https://www.astm.org/e2339-15.html (accessed on 1 April 2022).

45. Intel Corporation. Stylized as Intel, Is an American Multinational Corporation and Technology Company Headquartered in Santa Clara, California, USA. Available online: https://www.intel.com/ (accessed on 1 April 2022).
46. Nvidia. Nvidia Corporation Is an American multinational Technology Company Incorporated in Delaware and Based in Santa Clara, California, USA. Available online: https://www.nvidia.com/ (accessed on 1 April 2022).

Article

A Novel Sub-Pixel-Shift-Based High-Resolution X-ray Flat Panel Detector

Jiayin Liu and Jae Ho Kim *

Image and A. I. Laboratory, Department of Electronics Engineering, Pusan National University, Busan 46241, Korea; liujiayinpnu@gmail.com
* Correspondence: jhkim@pusan.ac.kr; Tel.: +82-10-4042-2450

Abstract: In this paper, we describe a novel sub-pixel shift (SPS)-based X-ray flat panel detector (FPD), which can achieve high resolution while maintaining a high SNR (signal-to-noise ratio). In the proposed architecture, an XY precision shift stage is applied to complete the sub-pixel shift process. In addition, image acquisition and high-resolution image composition are integrated in the FPD hardware. According to the relevant standards for detector image quality evaluation, we tested and evaluated some image quality indicators. The results show that the proposed FPD with SPS outperforms the original FPD without SPS technology. More specifically, the measured pixel size of the proposed FPD was reduced from 162 to 140 µm for 2 × 2 sub-pixel shift mode, and 132 µm for 4 × 4 sub-pixel shift mode, that is, the basic spatial detector resolution was improved by 13.6% for the simplest 2 × 2 sub-pixel shift mode, and by 18.5% for 4 × 4 sub-pixel shift mode. With this method, a lower-price FPD is elevated both in resolution and SNR_n to meet imaging quality requirements.

Keywords: X-ray flat panel detector; high resolution; sub-pixel shift

Citation: Liu, J.; Kim, J.H. A Novel Sub-Pixel-Shift-Based High-Resolution X-ray Flat Panel Detector. *Coatings* **2022**, *12*, 921. https://doi.org/10.3390/coatings12070921

Academic Editor: Jianzhong Zhang

Received: 2 June 2022
Accepted: 27 June 2022
Published: 29 June 2022

Publisher's Note: MDPI stays neutral with regard to jurisdictional claims in published maps and institutional affiliations.

Copyright: © 2022 by the authors. Licensee MDPI, Basel, Switzerland. This article is an open access article distributed under the terms and conditions of the Creative Commons Attribution (CC BY) license (https://creativecommons.org/licenses/by/4.0/).

1. Introduction

The X-ray flat panel detector (FPD) [1,2] is a photographic element used in digital radiography. In the same way that a normal digital camera uses a CMOS sensor to receive light passing through a lens and converts it into an image, a flat panel detector converts X-rays passing through the object into a digital image.

X-ray flat panel detectors have been widely used in security, industrial, and medical applications in place of conventional image intensifiers (I.I.) [2] and imaging plates (IP) [3]. The dynamic range of FPD is greater than that of I.I., and the images can be viewed in real time, without the need to remove the plate and extract the images, as with IP. The main applications of FPD in the industrial field are X-ray and CT non-destructive inspection, including casting and welding inspection, 3D printing inspection, SMT (surface-mounted technology) and semiconductor inspection, new energy battery inspection, security check, and so on; in the medical field, the main applications cover almost all X-ray equipment, including DR (digital radiography), DRF (dynamic DR), DM (digital mammography), CBCT (dental CT), DSA (digital subtraction angiography), C-arm X-ray systems, and so on.

The first flat-panel detector DR systems based on amorphous silicon [4] and amorphous selenium [5] were introduced in 1995. Subsequently, major medical imaging equipment companies conducted preliminary research on the technology. In the late 1990s, GE and Perkin Elmer in cooperation, Thales, Siemens, Philips in co-investment with Trixell, Varex, Canon Medical, and other companies developed amorphous silicon flat panel detectors [6]. Around 2010, amorphous silicon flat panel detector technology further proliferated, and traditional film giants Carestream, Fujifilm, Konica, and Agfa also developed flat panel detectors. Meanwhile, South Korea's Viewworks and Rayence and China's PZImaging, KangZhong, and i-Ray also launched their own amorphous silicon flat panel detectors [7].

Several kinds of FPDs have been developed over the past few decades [8]. Existing FPDs are divided into two types: indirect conversion detectors and direct conversion detectors [9]. The principle of indirect conversion FPDs is that X-ray irradiation is first converted to visible light through the scintillator; then, the digital image is read out using the principle of visible light cameras. Its basic structure includes the following: scintillator, sensor and readout circuit, and peripheral control circuit. The scintillator and sensor are the core part and determine the main performance of the FPD. Amorphous silicon (a-Si), CMOS, IGZO [10], and flexible FPD are all indirect conversion detectors. In contrast, direct conversion FPDs do not require scintillators. They convert X-rays into electrical signals directly after the X-rays are collected by the photoconductive semiconductor material. Therefore, the basic structure of the direct conversion FPD includes the following: sensor and readout circuit, peripheral control circuit, and the sensor (photoconductive semiconductor), which is the core part.

The technologies used by different types of detectors and their main application areas are shown in Table 1.

Table 1. The technologies used by different types of detectors and their main application areas.

Major Categories	Types	Detector Technology	Main Application Areas
Indirect conversion detectors	a-Si FPD	Scintillator + a-Si + TFT	DR/DRF, radiotherapy, industrial
	IGZO FPD	Scintillator + a-Si + IGZO	DR/DRF, intervention
	CMOS FPD	Scintillator + CMOS	Dental, mammograph, surgical, industrial
	flexible FPD	Scintillator + a-Si + TFT	Mobile healthcare
Direct conversion detectors	a-Se FPD	a-Se + TFT	Mammograph
	photon counting FPD	CdTe/CZT +CMOS	CT, breast CT

The two most important metrics to evaluate the imaging quality of FPDs are the signal-to-noise ratio (SNR) [11] and spatial resolution (SR) [12]. FPDs have different SNR and SR due to different materials, structures, and processes. SNR affects the ability to distinguish density differences in different tissues (i.e., high SNR means high density resolution), and SR affects the ability to distinguish fine spatial structures of tissues. To improve the spatial resolution, the pixel size of the detector needs to be made smaller. However, too small a pixel size leads to a decrease in SNR. As a direct consequence, the density resolution decreases, the image signal-to-noise ratio deteriorates, and the image becomes unusable. It is generally necessary to find a balance between these two imaging indexes. There is no FPD on the market that performs well in both SNR and spatial resolution.

In the medical field, the target of DR equipment is to observe and distinguish the density of different tissues in the chest and lungs. Therefore, the requirements for density resolution (SNR) are relatively high, while the pixel size of FPDs are larger, generally 139 μm, to easily obtain images with higher contrast. Accordingly, amorphous silicon FPDs are generally selected, which are more conducive to diagnosis. For the examination of extremity joints and breast, better imaging of structural spatial details is needed, so the pixel size of the FPD should be small, generally 50 to 76 μm, in order to obtain high-spatial-resolution images. Generally, amorphous selenium or CMOS FPDs are selected for these types of applications. The main reason for the higher requirement of spatial resolution in the industrial field than in the medical field is the continuous improvement of product quality control requirements. Defects that need to be detected are getting smaller and smaller as the result of the process upgrading. At present, film imaging is still used in a large number of industrial inspection scenarios, mostly on weld and casting inspection. In addition to regulatory factors, the main reason why users choose film over FDPs is that the actual spatial resolution of FPDs cannot reach the film imaging level. The industry expects that the spatial resolution could be further improved without decreasing SNR.

The resolution of the original image can be improved by pure software methods, which some people call super-resolution. Currently, there are three methods to achieve image super-resolution by software: interpolation-based methods, reconstruction-based methods, and learning-based methods.

Interpolation-based methods try to increase image resolution by filling in the corresponding pixel values on the empty spots after zooming in. However, these methods are efficient but ineffective [13,14]. Reconstruction-based methods align multiple low-resolution images of the same scene with sub-pixel accuracy on the space, obtaining the motion offsets between high- and low-resolution images and constructing the spatial motion parameters in the observation model to obtain a high-resolution image [15–19]. The core idea of reconstruction-based super-resolution methods is to trade temporal bandwidth (acquiring multiple image sequences of the same scene) for spatial resolution. With the significant development of deep learning technology, image super-resolution technology does have a wide range of practical applications in the fields of games, movies, medical imaging, and so on. Learning-based methods adopt the end-to-end mapping function of low-resolution images to high-resolution images by neural networks. Using the prior knowledge acquired by the model to obtain high-frequency details of the image, they are considered the best way to enhance image resolution at present. The mainstream algorithms are SRCNN, SRGAN, ESRGAN, and so on. Among them, the SRCNN method [20,21] has the simplest network structure, using only three convolutional layers, and the framework is flexible in choosing parameters and supports customization. The disadvantage is that the details are not sufficiently represented, and the results obtained are too smooth and unrealistic when the magnification exceeds four. The SRGAN method [22] generates realistic textures for a single image to complement the lost details but introduces high-frequency noise at the same time. The ESRGAN output image has better image quality with more realistic and natural textures, and it tops the PIRM2018-SR challenge [23]. In the experimental section, we will cite the results of ESRGAN as a part of the comparison data.

The spatial resolution of FPDs is mainly determined by the pixel size, and a smaller pixel size leads to a higher spatial resolution. However, as a result of various factors such as process, cost, imaging quality, and imaging field of view, the pixel size cannot be reduced indefinitely at the actual product level. The common pixel size of amorphous silicon detectors is generally 100, 139, and 200 μm. The actual measured spatial resolution is worse than the theoretical spatial resolution corresponding to the pixel size. Types of scintillators (CSI, GdOS, etc.), scintillator thickness, and vapor deposition process will affect the actual spatial resolution. The measured spatial resolution of the detector with a nominal pixel size of 139 μm is typically between 150 to 190 μm.

We have developed a novel high-resolution X-ray FPD based on sub-pixel shift (SPS) technology, which can improve the inherent spatial resolution of the detector without degrading the image quality. We designed and implemented the underlying hardware and software for sub-pixel acquisition based on an amorphous silicon glass substrate, and users can easily control the sub-pixel acquisition accuracy of the detector through commands. Experiments show that the measured pixel size of this new high-resolution detector is reduced from 162 to 132 μm, that is, the basic spatial resolution of the detector is improved by 18.5%.

2. Evaluation Metrics

2.1. Basic Spatial Detector Resolution

"ISO-19232-5-2018, Non-destructive testing—Image quality of radiographs—Part 5: Determination of the image unsharpness and basic spatial resolution value using duplex wiretype image quality indicators" specifies a method of determining the total image unsharpness and basic spatial resolution of radiographs and radioscopic images [12]. The most important metric in this standard relevant to this work is SR_b^{image} (basic spatial detector resolution), which is determined from the smallest number of the duplex wire pair.

The basic spatial detector resolution is measured with the duplex wire IQI (image quality indicator) directly placed on the detector without object.

The duplex wire IQI with up to 13 wire pairs can be used effectively with tube voltages up to 600 kV, and the specification of duplex wire-type IQI can be seen in Figure 1.

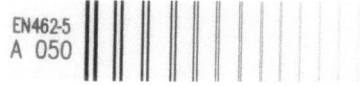

Specification:
- Duplex IQI consists of 13 or 15 wire pairs from 1D to 13D or 15D
- material for wires 1D to 3D is tungsten and for wires 4D to 15D is platinum
- distance between wires of each wire pair equals exactly diameter of wires
- wires are casted in a transparent, resistant and dimensional stable plastic
- standard and serial number are indelible casted and shown on each image
- design-type test for Duplex IQI is in process at BAM / Berlin

Figure 1. Specification of duplex wire-type IQI.

Measurement and calculation procedure for $SR_b^{detector}$: The duplex wire IQI should be placed directly on the detector with an angle between 2° and 5° to the rows/columns of the detector. No image processing shall be used other than gain/offset and bad pixel corrections.

If digital images are evaluated with a profile function, the element with the smallest wire number of the duplex wire pair, which is separable by a profile function with less than 20% modulation depth, is taken as the limit of discernibility for digital radiography. The profile function shall be evaluated from linearized pixel profiles. The measurement shall be done with the profile function of image-processing software across the middle area of the IQI image, integrating along the wires of about 30% to 60% of the duplex wires' length in order to obtain a robust repeatable value, but shall use a minimum of an 11-pixel width line profile to avoid variability along the length of the wires.

2.2. Image Quality Evaluation

"EN 12681-2:2017 Founding—Radiographic testing, part 2: Techniques with digital detectors" [24] specifies the recommended procedure for detector selection and radiographic practice, and the requirements for digital radiographic testing by either computed radiography (CR) or radiography with digital detector arrays (DDA) of castings. Three metrics are selected for the image quality evaluation of the proposed high-resolution X-ray flat panel detector: contrast sensitivity, SR_b^{image} (basic spatial resolution of a digital image), and SNR_n (normalized signal-to-noise ratio).

2.2.1. Contrast Sensitivity

Unless otherwise agreed, the contrast sensitivity of digital images shall be verified by use of IQIs, in accordance with EN ISO 19232-1 or EN ISO 19232-2. A single wire IQI is shown in Figure 2.

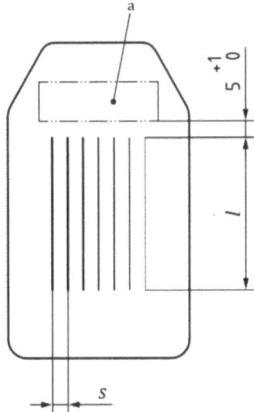

Figure 2. A schematic diagram of single wire IQI, in which, l: length of the wires, s: wire centerline spacing, a: space for identification marking.

The single wire IQI system is based on a series of 19 wires of different diameters, which are specified in Table 2 together with the relevant tolerances and the wire numbers. This series of wires has been subdivided into four overlapping ranges of seven consecutive wire numbers: W1 to W7, W6 to W12, W10 to W16, and W13 to W19. The seven wires in an IQI are arranged parallel to each other. The lengths of the wires (l) are 10, 25, or 50 mm. The single wire IQI shall be placed on the source side of the test object.

Table 2. Minimum image quality requirements for the visibility of wire IQIs for class A or B.

Penetrated Thickness (mm)		Minimum Wire IQI Value for Class A	Minimum Wire IQI Value for Class B
Lower Thickness Limit	Upper Thickness Limit	IQI at Source Side	IQI at Source Side
-	1.2	W18	-
>1.2 for Class A and 0 for Class B	2	W17	W18
>2	3.5	W16	W17
>3.5	5	W15	W16
>5	7	W14	W15
>7	10	W13	W14
>10	15	W12	W13
>15	25	W11	W12
>25	32	W10	W11
>32	40	W9	W10
>40	55	W8	W9
>55	85	W7	W8
>85	150	W6	W7
>150	200	W5	W6
>200	250	W4	W5
>250	380	W3	W4
>380	-	W2	W3

2.2.2. Basic Spatial Resolution of a Digital Image

SR_b^{image} (basic spatial resolution of image) corresponds to the effective pixel size and indicates the smallest geometrical detail that can be resolved in a digital image. For this measurement, the duplex wire IQI should be placed on the object (source side).

2.2.3. Normalized Signal-to-Noise Ratio

SNR (signal-to-noise ratio) is the ratio of mean value of the linearized gray values to the standard deviation of the linearized gray values (noise) in a given region of interest in a digital image. SNR_N (normalized signal-to-noise ratio) is normalized by the basic spatial resolution SR_b^{image} as measured directly in the digital image and/or calculated from measured $SNR_{measured}$.

$$SNR_N = SNR_{measured} \times \left(88.6 \text{ μm}/SR_b^{image}\right), \quad (1)$$

2.2.4. Minimum Image Quality Values

The radiographic techniques for film replacement are divided into two classes in EN 12681-2:2017:
- Class A: basic techniques
- Class B: improved techniques

Tables 2–4 show the minimum image quality requirements accordingly.

Table 3. Minimum image quality requirements for the visibility of duplex wire IQIs for class A or B.

Penetrated Thickness (mm)		Minimum DW Value for Class A IQI at Source Side	Minimum DW Value for Class B IQI at Source Side
Lower Thickness Limit	Upper Thickness Limit		
-	2	D12	D13+
>2	5	D10	D13
>5	10	D9	D12
>10	24	D8	D11
>24	40	D7	D10
>40	55	D7	D9
>55	85	D6	D9
>85	150	D6	D8
>150	200	D5	D8
>200	250	D5	D7
>250	380	D4	D7
>380	150	D4	D6

Table 4. Minimum SNR_N values for the digital radiography of aluminum, magnesium, and zinc.

Radiation Source	Minimum SNR_N for Class A	Minimum SNR_N for Class B
≤150 kV	70	120
150 to 250 kV	70	100
250 to 500 kV	70	100

3. High-Resolution Flat Panel Detector Design

3.1. The Proposed Image Detector Architecture

An indirect detector contains a layer of scintillating material that converts the X-rays into visible photons (light). Behind the scintillator, an array of photodiodes converts the light into an electrical signal. The array of photodiode pixels is similar in concept to a camera's image sensor: a high density of pixels creates a high-resolution image in which small features are clearly and sharply rendered. The stored charge of each pixel is proportional to the intensity of the incident X-rays. Under the action of the control circuit, the stored charge of each pixel is scanned and read out, and the digital signal is output after A/D conversion and transmitted to the computer for image processing to form a digital X-ray image. A based architecture demonstration of the indirect FPD is illustrated in Figure 3 [25].

The entire architecture of the proposed novel detector is shown in Figure 4, using CSI scintillator as the sensor; a-Si-based TFT as the converter; data operation part with integrated read, select, and pack; and a main state machine to control multiple modules and operation modes. Command passing through cmd process, buffer, and synchronization are designed. Udp/ip protocol stack and 1000 M wired network are used for all the command and data transfer.

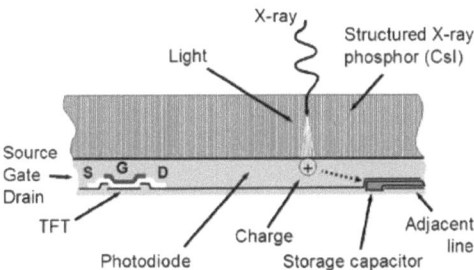

Figure 3. A based architecture demonstration of the indirect FPD.

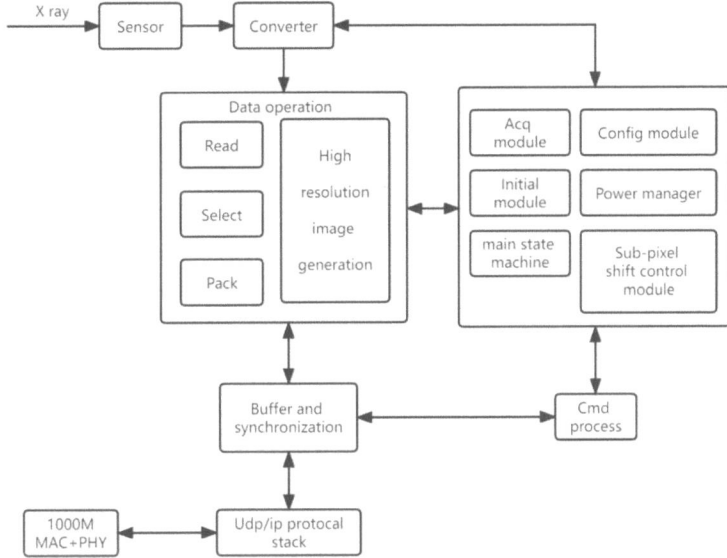

Figure 4. The entire architecture of our novel detector.

A circuit design diagram of some basic hardware modules of the detector is shown in Figure 5, which includes: (a) AD chip interface, (b) gate driver chip interface, (c) network interface, and (d) main control board.

3.2. Sub-Pixel Shift Design

Due to process and other constraints, the pixel size of the detector panel is limited. We use a sub-pixel shift design that allows each pixel to be spatially displaced in a controlled and precise sub-pixel scale in the XY direction, thus realizing pixel interpolation at the physical hardware level to improve the spatial resolution of the detector.

The procedure of sub-pixel shift is illustrated in Figure 6. It shows the four acquired original images and the high-resolution image composed.

In Figure 6, the resolution of the original image is $S \times S$. Assuming that the inherent pixel size of the detector is L, then the distance between each two adjacent pixel centroids is L as well. Divide L equally into N parts to get the step size of each precise displacement. image a is the original image acquired at the original position (0, 0), image b is the original image acquired after moving $L/2$ distance to the right (X direction), image c is the original image acquired after moving $L/2$ distance to the bottom (Y direction), and image d is the original image acquired after moving $L/2$ distance to the right (X direction) and to the bottom (Y direction). Image e is the generated higher-resolution image with a resolution of $2S \times 2S$. In practice, when 2×2 mode is selected, first of all, the memory space inside the detector is initialized with 4 times size of the original required, and all the pixel values are set to zero. Then, the pixel values of the four individual images are realigned to create a high-resolution image, as illustrated in Figure 6.

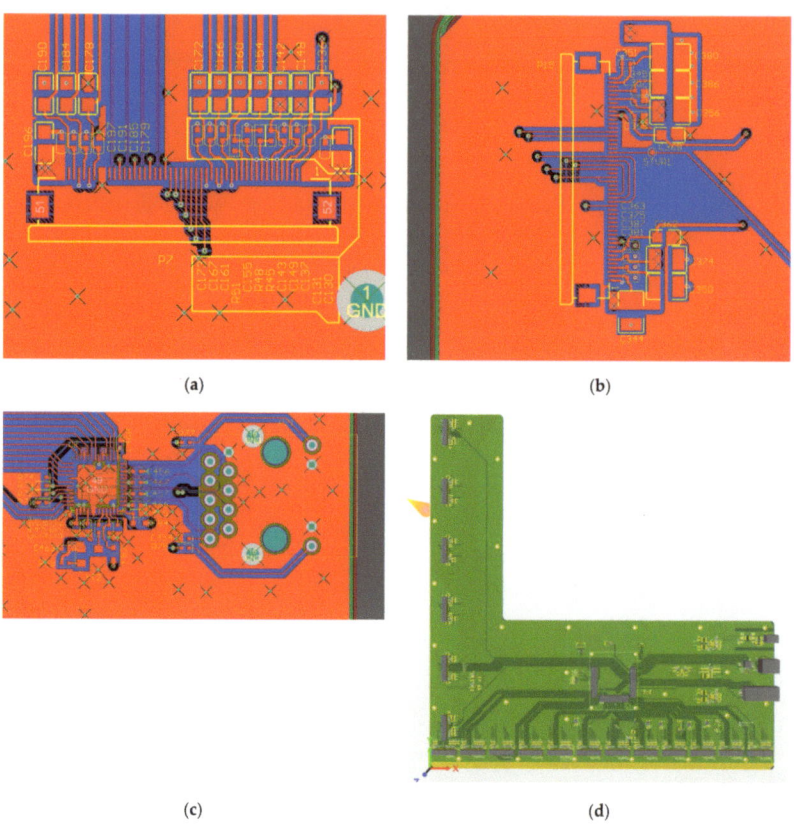

Figure 5. The circuit design diagram of some basic hardware modules of the detector: (**a**) AD chip interface, (**b**) gate driver chip interface, (**c**) network interface, and (**d**) main control board.

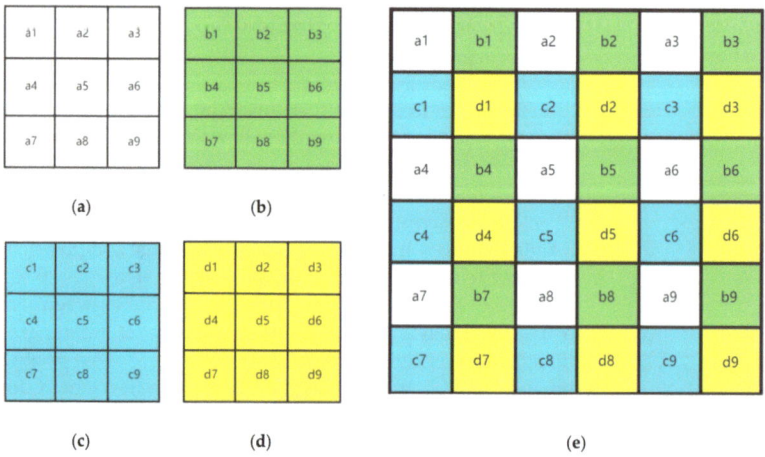

Figure 6. How high-resolution images are created by combining the pixels of individual native-resolution images. (**a**–**d**) are four acquired original images, and (**e**) is 2 × 2 mode higher-resolution image.

For modes other than 2 × 2, the operation is similar. We found in the actual test that the improvement of spatial resolution is limited after N > 4.

At the hardware level, we use an XY precision shift stage to fix the detector panel on it and complete the sub-pixel shift process. Image acquisition and image stitching are realized through integrated control and acquisition commands. The precision shift stage is shown in Figure 7. It uses noncontact optical encoders to measure the position directly at the platform with the greatest accuracy and achieves unidirectional repeatability to 0.05 µm and incremental linear encoder with 1 nm resolution. There are a variety of acquisition modes integrated in the main control board. In the application side, through the commands, the users can directly select different modes of precision for the high-resolution acquisition. Now, it supports a total of four modes: 1 × 1, 2 × 2, 3 × 3, and 4 × 4. The core parameters of the proposed sub-pixel shift FPD are shown in Table 5.

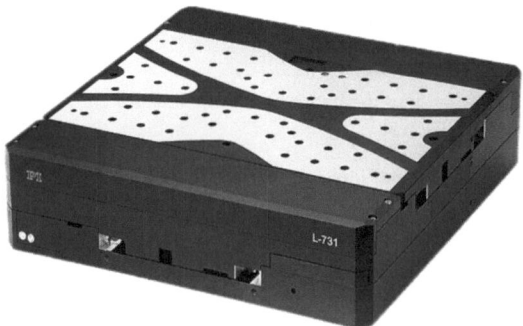

Figure 7. The XY precision shift stage.

Table 5. The core parameters of the proposed sub-pixel shift FPD.

Parameter	Value
Type	a-Si
Scintillator	CSI
Number of rows	3072
Number of columns	3072
Pixel pitch (µm)	139
Imaging area (mm^2)	430 × 430
A/D bit	16
Frame per second (fps)	6 (1 × 1), 20 (2 × 2)
Sub-pixel shift modes	1 × 1, 2 × 2, 3 × 3, 4 × 4

4. Experiments

4.1. Experimental Environment

We have cooperated with Deepsea Precision Co., Ltd. [26], a manufacturer specializing in X-ray inspection equipment, to test the FPD and collect all the images from an actual X-ray environment. Figure 8 is a piece of industrial X-ray inspection equipment from Deepsea Precision Co., Ltd., with our FPD integrated into this testing system.

The basic configuration of the core imaging chain components from this typical X-ray inspection equipment of Figure 8 is shown in Table 6.

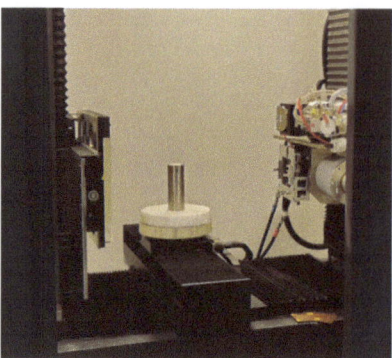

Figure 8. The industrial X-ray inspection equipment from Deepsea Precision Co., Ltd.

Table 6. The basic configuration of the core imaging chain components from the X-ray inspection equipment of Figure 8.

Device Name	Brand/Model	Basic Configuration
X-ray emission device (macro-focus)	Gulmay/CF500	500 kV, focus size 0.4/1.0 mm
X-ray receiver device (flat panel detector)	Our sub-pixel shift FPD	430 mm × 430 mm, pixel size 139 μm
Workstation software	Deepsea/DeepVISION	GPU-based architecture

4.2. Experimental Data

4.2.1. Basic Spatial Detector Resolution Testing

The basic spatial detector resolution $SR_b^{detector}$ was tested according to ISO-19232-5-2018 under the following conditions: SDD (source detector distance) = 1000 mm, voltage = 90 kV, current = 1.3 mA. In our experiments, a software named DeepVision was used, and least squares curve fitting was adopted to calculate the SRb value in this software. The measuring procedure was carried out according to ISO-19232-5. We improved measurement precision by averaging the testing values of repeated measurements. Test results are shown in Table 7, and more detailed test images and data are shown in Figure 9. The results show that the basic spatial resolution of the novel detector proposed in this paper is greatly improved. The basic spatial detector resolution of the original image without any sub-pixel shift is 162 μm, the basic spatial detector resolution of the image after 2 × 2 sub-pixel shift is 140 μm, and the basic spatial detector resolution of the image after 4 × 4 sub-pixel shift is 132 μm. To summarize, the basic spatial detector resolution was improved by 13.6% after 2 × 2 shift and 18.5% after 4 × 4 sub-pixel shift.

Table 7. Test results of the basic spatial detector resolution $SR_b^{detector}$.

The Basic Spatial Resolution of the Detector	Original Image	2 × 2 Sub-Pixel Shift Image	4 × 4 Sub-Pixel Shift Image
$SR_b^{detector}$	162 μm (D8)	140 μm (D9)	132 μm (D9)

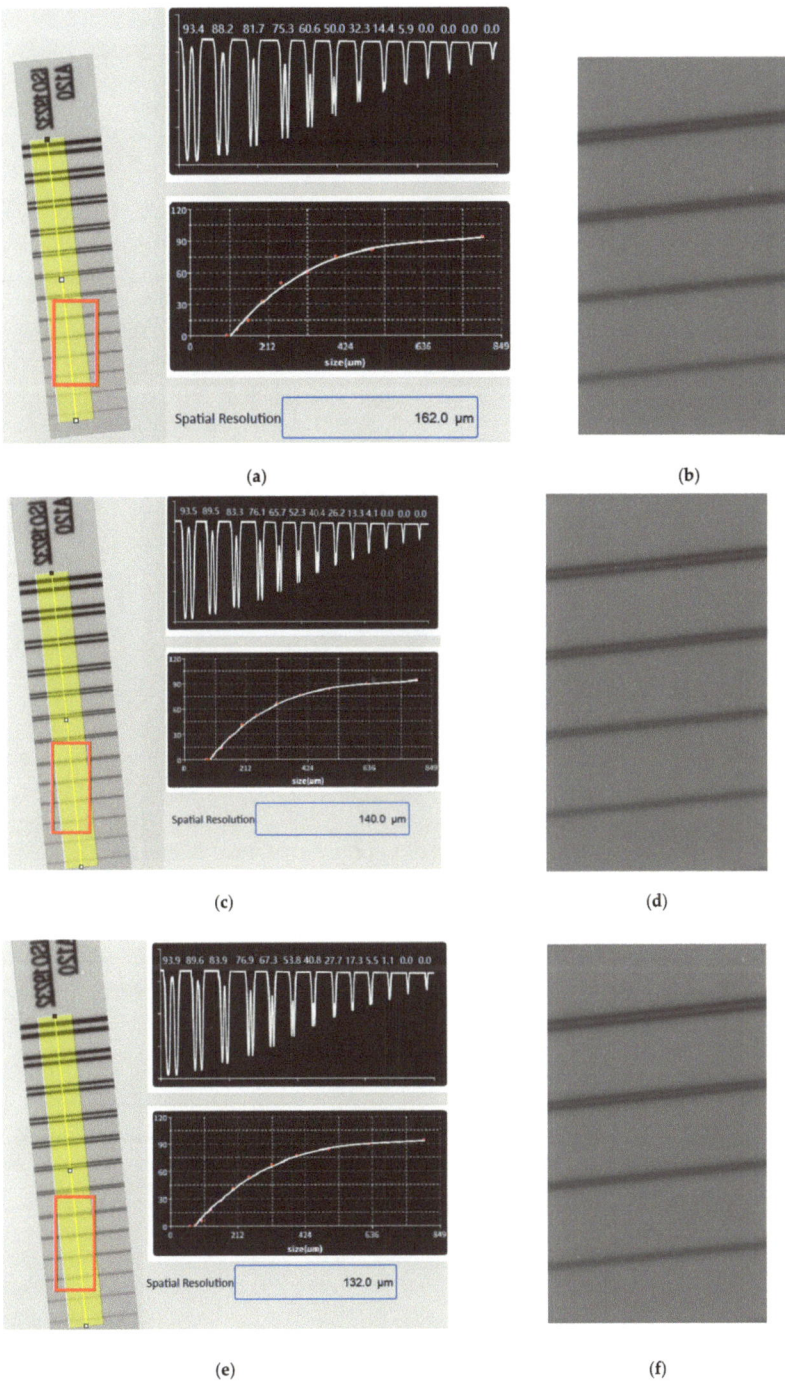

Figure 9. The basic spatial detector resolution $SR_b^{detector}$: (**a**) original image, (**c**) 2 × 2 sub-pixel shift image, (**e**) 4 × 4 sub-pixel shift image; (**b**,**d**,**f**) are the corresponding local enlargement of the boxed areas of the duplex wire IQI.

4.2.2. Image Quality Testing

According to the standard EN 12681-2:2017, the duplex wire IQI and single wire IQI are fixed on a step wedge for image quality testing. The step wedge is aluminum, and the thinnest layer is 10, then 20, 40 mm, and so on. The testing was conducted at 20 mm thickness. The test condition is: SDD (source detector distance) = 1000 mm, voltage = 140 kV, current = 1.5 mA. A total of three tests are conducted: single wire IQI, duplex wire IQI, and SNR/SNR_n.

Test 1: Single Wire IQI Result

The test results are shown in Figure 10. The visibility levels of the single wire IQI on the original image and the resolution-enhanced image after sub-pixel shift are the same, both being W14. This result indicates that the proposed sub-pixel-shift-based method does not cause a decrease in contrast sensitivity.

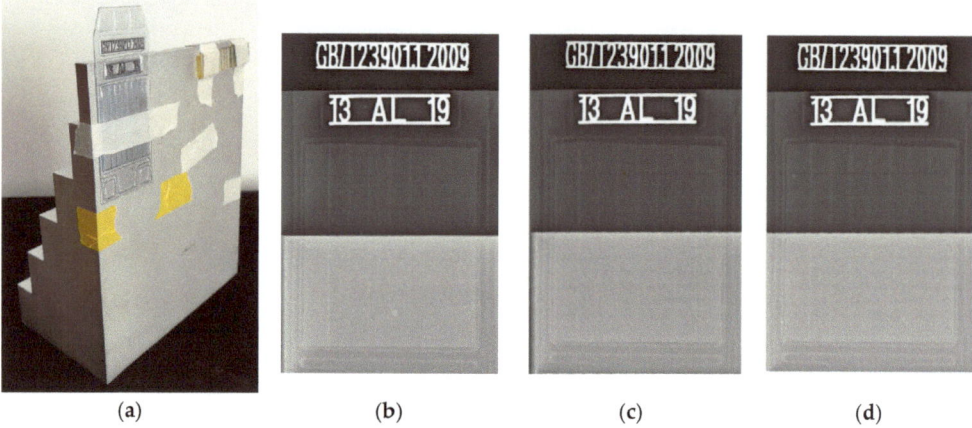

Figure 10. Test results of the contrast sensitivity: (**a**) experimental device, (**b**) original image, (**c**) 2 × 2 Sub-Pixel Shift Image, (**d**) 4 × 4 Sub-Pixel Shift Image.

Test 2: Duplex Wire IQI Result

For the image basic spatial resolution test, the duplex wire IQI should be attached to the step wedge with a certain magnification to evaluate the spatial resolution performance when imaging the actual object. Test results are shown in Table 8, and more detailed test images and data are shown in Figure 11.

Table 8. Test results of the image spatial resolution SR_b^{image} of the detector.

The Basic Spatial Resolution of the Detector	Original Image	2 × 2 Sub-Pixel Shift Image	4 × 4 Sub-Pixel Shift Image
SR_b^{image}	159 μm (D8)	137 μm (D9)	132 μm (D9)

The results show that the image spatial resolution of the novel detector proposed in this paper is greatly improved. The image spatial resolution of the original image without any sub-pixel shift is 159 μm (D8), which does not meet the minimum image requirement according to Table 3. In contrary, the image spatial resolution of the image after 2 × 2 sub-pixel shift is 137 μm, and the image spatial resolution of the image after 4 × 4 sub-pixel shift is 132 μm. That is to say, the image quality achieved by the proposed sub-pixel shift method has been improved to meet the requirements for Class A.

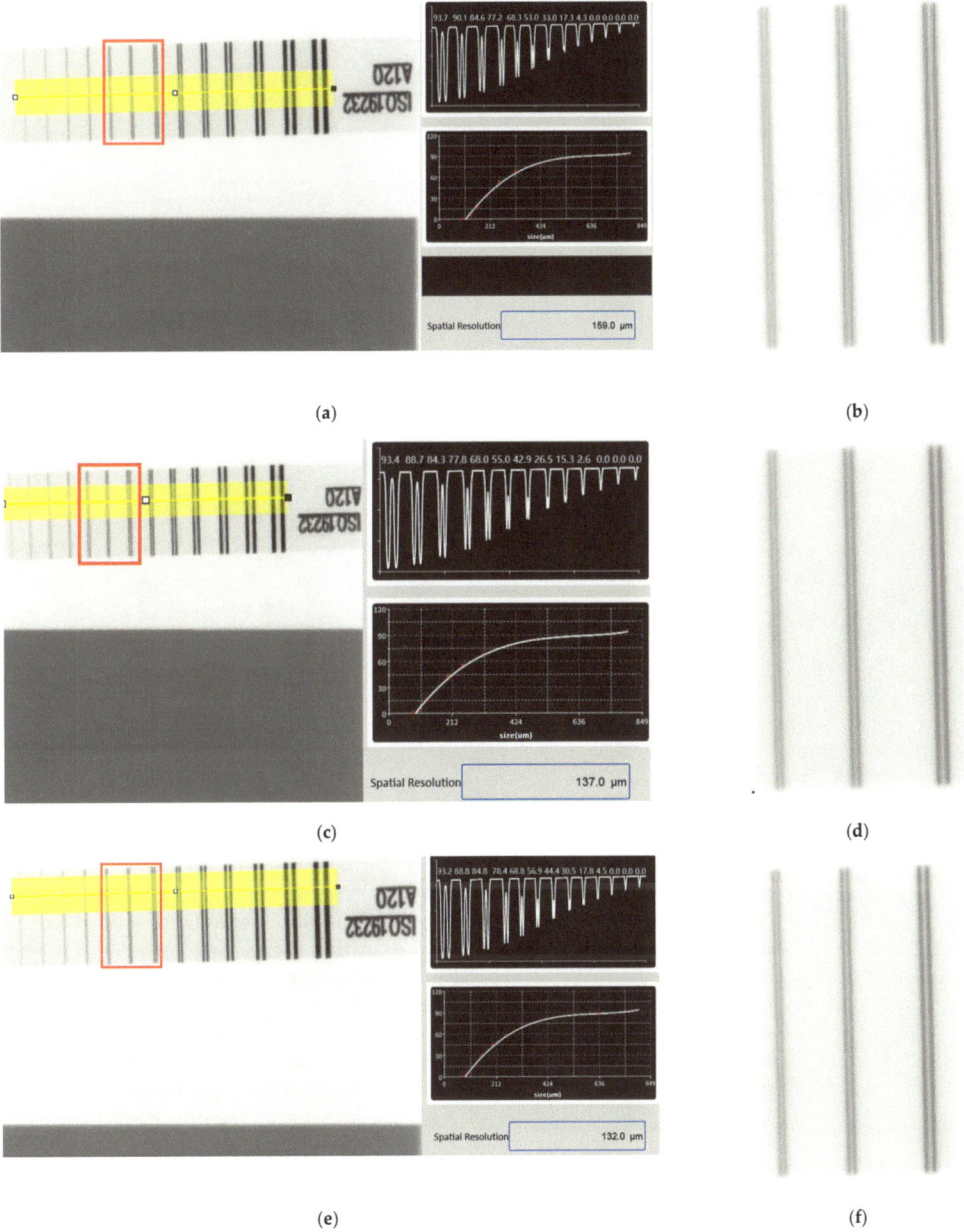

Figure 11. The image spatial resolution SR_b^{image} of the detector: (**a**) original image, (**c**) 2 × 2 sub-pixel shift image, (**e**) 4 × 4 sub-pixel shift image; (**b,d,f**) are the corresponding local enlargement images of the boxed areas of the duplex wire IQI.

Test 3: SNR/SNR_n Result

According to the standard, SNR/SNR_n is calculated in different grayscale distribution areas, and an SNR/SNR_n graph is generated, as shown in Figure 12 with the average grayscale value of the selected area in the horizontal coordinate and the SNR/SNR_n value in the vertical coordinate. It can be concluded that the proposed sub-pixel-shift-based high-resolution FPD in this paper improves slightly in SNR and significantly in SNR_n due to the great improvement in spatial resolution.

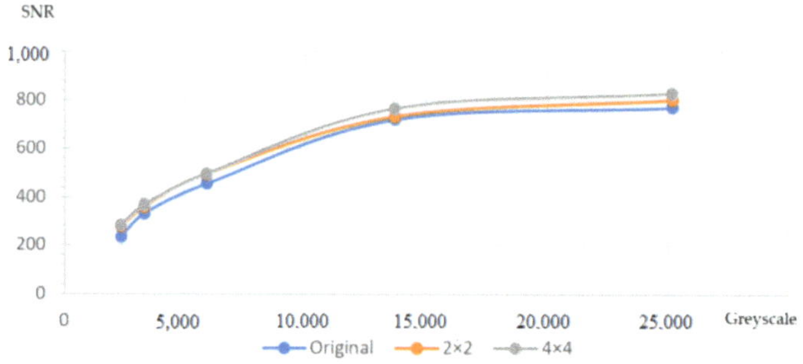

Figure 12. The SNR/SNR_n of the detector: (**a**) SNR, (**b**) SNR_n.

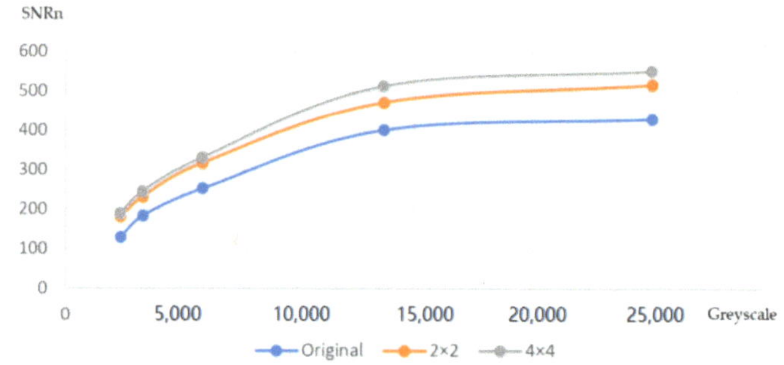

4.2.3. Comparison with Software Interpolation Methods

Some of the software-based super-resolution methods have been discussed previously, among which the ESRGAN algorithm is currently considered to be the most effective method. We compared the results of the ESRGAN algorithm with ours, as shown in Figure 13. ESRGAN improves the overall spatial resolution, but introduces local artifacts and noise, which can seriously affect the inspection results and is therefore not applicable. The arrows indicate artifacts produced by ESRGAN algorithm. In the proposed method, an XY precision shift stage is adopted to move the detector. The FPD frame rate is 6 FPS, which means 6 frames per second. The shift stage moves the detector to the next position during each image readout time. In other words, the movement of detector does not occupy extra time. In general, for 2×2 mode, the total time for generating a high-resolution image is less than 1 s. For 4×4 mode, the total time is about 3 s. For ESRGAN method, the total time for upsampling a 3072×3072 image to 6144×6144 is about 800 ms in our experiments.

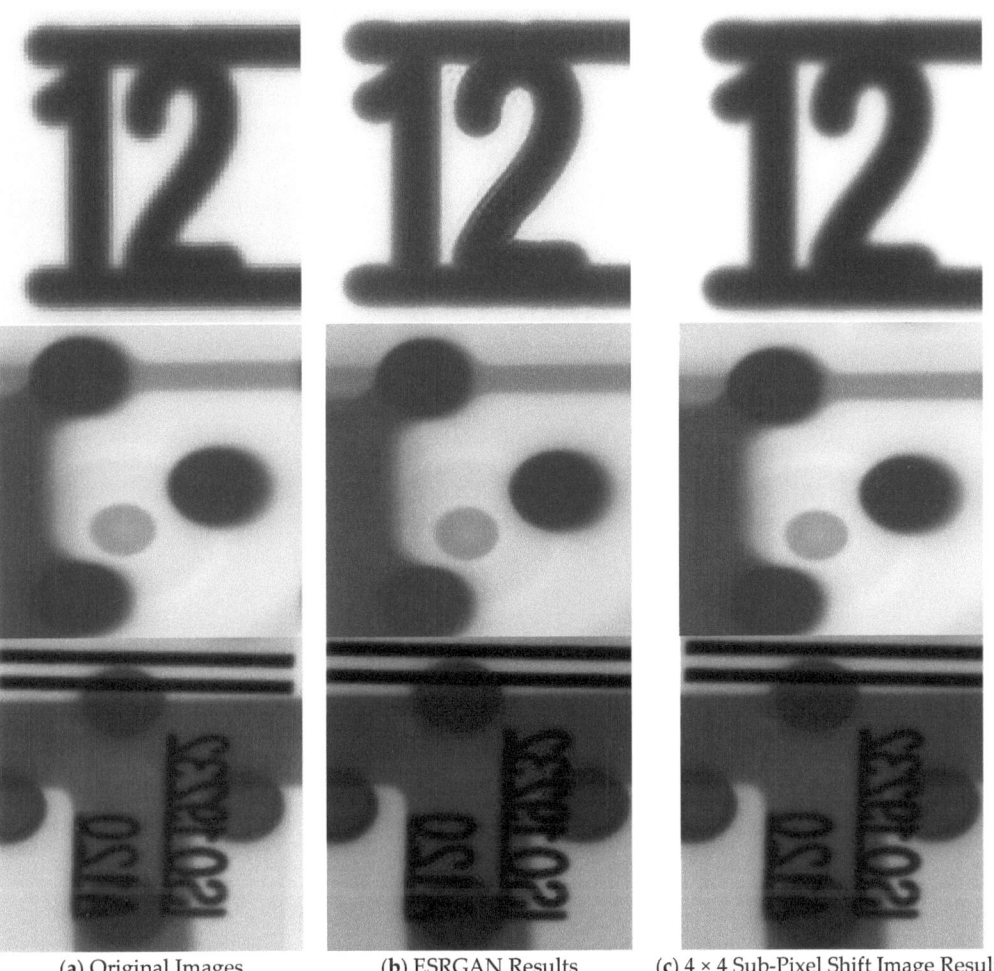

(a) Original Images (b) ESRGAN Results (c) 4 × 4 Sub-Pixel Shift Image Results

Figure 13. The results of the ESRGAN algorithm and ours on real object images. ESRGAN improves the overall spatial resolution but introduces local artifacts and noise. The arrows indicate artifacts produced by ESRGAN algorithm.

5. Conclusions

In this paper, we describe a novel sub-pixel shift (SPS)-based X-ray flat panel detector (FPD) that can achieve high resolution while maintaining high SNR. In comparison with the existing techniques, the proposed high-resolution FPD has the following features: (i) a sub-pixel-shift-based acquisition and data composition strategy are integrated in the detector; (ii) the pixel size of the detector was reduced from 162 to 132 μm, that is, the basic spatial detector resolution was improved by 13.6% in the simplest 2 × 2 sub-pixel shift mode, and by 18.5% in 4 × 4 sub-pixel shift mode; (iii) for X-ray images, the test results show a great improvement in image spatial resolution, as well as high SNR_n.

The main goal of X-ray image processing is to avoid introducing artifacts. This is especially important for X-ray non-destructive testing. Image super-resolution by software algorithms, including interpolation methods, reconstruction-based methods, and deep learning methods, have been proposed and studied for many years. Some of them can achieve great performance on visible light images. However, they have poor performance on X-ray images with high dynamic range and internal noise. Some may cause severe image distortion and artifacts, while others may enlarge the image noise. The poor performance of these high-resolution methods might lead to false alarms for X-ray non-destructive testing. The proposed method has been applied to the testing system, and the performance has been demonstrated by experiments. The results show that our method is effective. Both spatial resolution and SNRn have been improved without introducing artifacts.

In the future, we plan to apply the methodology in this work to other types of X-ray detectors. In addition, we will try to increase the frame rate of the detector using a 10 Gigabit network or fiber optic interface, which allows the sub-pixel acquisition process to be done faster.

Author Contributions: Conceptualization, J.L. and J.H.K.; methodology, J.L.; software, J.L.; validation, J.L.; formal analysis, J.L.; investigation, J.L. and J.H.K.; resources, J.L.; data curation, J.L.; writing—original draft preparation, J.L.; writing—review and editing, J.H.K.; visualization, J.L.; supervision, J.H.K.; project administration, J.L.; funding acquisition, J.L. All authors have read and agreed to the published version of the manuscript.

Funding: This research was supported by Guangdong Science and Technology Project, Grant Nos. 2020A0505100012, 2017B020210005, and BK21FOUR; and Creative Human Resource Education and Research Programs for ICT Convergence in the 4th Industrial Revolution.

Institutional Review Board Statement: Not applicable.

Informed Consent Statement: Not applicable.

Data Availability Statement: Not applicable.

Acknowledgments: Thanks to Deepsea Precision Co., Ltd. and for their X-ray inspection equipment and support for the FPD testing. Thanks to all the people who contributed to the FPD and image testing. Special thanks to Stan, a colleague of J.L., who did a lot of work in building the testing environment.

Conflicts of Interest: The authors declare no conflict of interest.

References

1. Rowlands, J.A.; Yorkston, J. *Handbook of Medical Imaging*; Spie Press: Bellingham, WA, USA, 2000.
2. Baba, R.; Konno, Y.; Ueda, K.; Ikeda, S. Comparison of flat-panel detector and image-intensifier detector for cone-beam CT. *Comput. Med. Imaging Graph.* **2002**, *26*, 153–158. [CrossRef]
3. Travish, G.; Rangel, F.J.; Evans, M.A.; Hollister, B.; Schmiedehausen, K. Addressable flat-panel X-ray sources for medical, security, and industrial applications. In Proceedings of the Advances in X-ray/EUV Optics and Components VII, San Diego, CA, USA, 12–16 August 2012; Volume 8502, p. 85020L. [CrossRef]
4. Antonuk, L.E.; Yorkston, J.; Huang, W.; Siewerdsen, J.H.; Boudry, J.M.; El-Mohri, Y.; Marx, M.V. A real-time, flat-panel, amorphous silicon, digital X-ray imager. *RadioGraphics* **1995**, *15*, 993–1000. [CrossRef] [PubMed]
5. Zhao, W.; Rowlands, J.A. X-ray imaging using amorphous selenium: Feasibility of a flat panel self-scanned detector for digital radiology. *Med. Phys.* **1995**, *22*, 1595–1604. [CrossRef] [PubMed]
6. Kump, K.; Grantors, P.; Pla, F.; Gobert, P. Digital X-ray detector technology. *RBM-News* **1998**, *20*, 221–226. [CrossRef]
7. Kotter, E.; Langer, M. Digital radiography with large-area flat-panel detectors. *Eur. Radiol.* **2002**, *12*, 2562–2570. [CrossRef] [PubMed]
8. X-ray Detectors for Medical, Industrial and Security Applications, YOLE. 2019. Available online: www.yole.fr (accessed on 10 March 2007).
9. Schumacher, D.; Zscherpel, U.; Ewert, U. Photon Counting and Energy Discriminating X-Ray Detectors-Benefits and Applications. In Proceedings of the 19th World Conference on Non-Destructive Testing 2016, Munich, Germany, 13–17 June 2016.
10. Freestone, S.; Weisfield, R.; Tognina, C.; Job, I.; Colbeth, R.E. Analysis of a new indium gallium zinc oxide (IGZO) detector. In Proceedings of the Medical Imaging 2000: Physics of Medical Imaging, Houston, TX, USA, 16–19 February 2020; Volume 11312, p. 113123W. [CrossRef]
11. Granfors, P.R.; Aufrichtig, R. DQE(f) of an amorphous-silicon flat-panel X-ray detector: Detector parameter influences and measurement methodology. In Proceedings of the Medical Imaging 2000: Physics of Medical Imaging, Houston, TX, USA, 16–19 February 2000; Volume 3977, pp. 2–13. [CrossRef]
12. *ISO-19232-5-2018*; Non-Destructive Testing—Image Quality of Radiographs—Part 5: Determination of the Image Unsharpness and Basic Spatial Resolution Value Using Duplex Wiretype Image Quality Indicators. International Organization for Standardization: Geneva, Switzerland, 2018.

13. Keys, R. Cubic convolution interpolation for digital image processing. *IEEE Trans. Acoust. Speech Signal Process.* **1981**, *29*, 1153–1160. [CrossRef]
14. Duchon, C.E. Lanczos filtering in one and two dimensions. *J. Appl. Meteorol.* **1979**, *18*, 1016–1022. [CrossRef]
15. Aly, H.; Dubois, E. Image up-sampling using total-variation regularization with a new observation model. *IEEE Trans. Image Process.* **2005**, *14*, 1647–1659. [CrossRef] [PubMed]
16. Marquina, A.; Osher, S.J. Image Super-Resolution by TV-Regularization and Bregman Iteration. *J. Sci. Comput.* **2008**, *37*, 367–382. [CrossRef]
17. Dai, S.; Han, M.; Xu, W.; Wu, Y.; Gong, Y.; Katsaggelos, A. SoftCuts: A Soft Edge Smoothness Prior for Color Image Super-Resolution. *IEEE Trans. Image Process.* **2009**, *18*, 969–981. [CrossRef] [PubMed]
18. Sun, J.; Xu, Z.; Shum, H.-Y. Image super-resolution using gradient profile prior. In Proceedings of the IEEE Conference on Computer Vision and Pattern Recognition, Anchorage, AK, USA, 23–28 June 2008; pp. 1–8.
19. Yan, Q.; Xu, Y.; Yang, X.; Nguyen, T.Q. Single Image Superresolution Based on Gradient Profile Sharpness. *IEEE Trans. Image Process.* **2015**, *24*, 3187–3202. [CrossRef] [PubMed]
20. Dong, C.; Loy, C.C.; He, K.; Tang, X. Learning a Deep Convolutional Network for Image Super-Resolution. In Proceedings of the ECCV 2014, Zurich, Switzerland, 6–12 September 2014. [CrossRef]
21. Dong, C.; Loy, C.C.; He, K.; Tang, X. Image super-resolution using deep convolutional networks. *IEEE Trans. Pattern Anal. Mach. Intell.* **2016**, *38*, 295–307. [CrossRef] [PubMed]
22. Ledig, C.; Theis, L.; Huszár, F.; Caballero, J.; Cunningham, A.; Acosta, A.; Aitken, A.P.; Tejani, A.; Totz, J.; Wang, Z.; et al. Photo-Realistic Single Image Super-Resolution Using a Generative Adversarial Network. In Proceedings of the 2017 IEEE Conference on Computer Vision and Pattern Recognition (CVPR), Honolulu, HI, USA, 21–26 July 2017.
23. Wang, X.; Yu, K.; Wu, S.; Gu, J.; Liu, Y.; Dong, C.; Qiao, Y.; Loy, C.C. ESRGAN: Enhanced Super-Resolution Generative Adversarial Networks. In Proceedings of the 15th European Conference on Computer Vision, ECCV 2018, Munich, Germany, 8–14 September 2018; pp. 63–79.
24. *EN 12681-2:2017*; Founding—Radiographic Testing, Part 2: Techniques with Digital Detectors. The Slovenian Institute for Standardization: Ljubljana, Slovenia, 2018.
25. Seibert, J.A. Flat-panel detectors: How much better are they? *Pediatr. Radiol.* **2006**, *36*, 173. [CrossRef] [PubMed]
26. Deepsea Precision Co., Ltd. Shenzhen City, China. Available online: http://www.shenhaijingmi.com/ (accessed on 15 May 2022).

Article

A Semi-Supervised Inspection Approach of Textured Surface Defects under Limited Labeled Samples

Yu He [1,*], Xin Wen [1] and Jing Xu [2]

1 Department of Software Engineering, Shenyang University of Technology, Shenyang 110870, China
2 Mechanical Engineering, Shenyang University of Technology, Shenyang 110870, China
* Correspondence: heyu_142616@sut.edu.cn; Tel.: +86-186-0243-6728

Abstract: Defect inspection is a key step in guaranteeing the surface quality of industrial products. Based on deep learning (DL) techniques, related methods are highly effective in defect classification tasks via a supervision process. However, collecting and labeling many defect samples are usually harsh and time-consuming processes, limiting the application of these supervised classifiers on various textured surfaces. This study proposes a semi-supervised framework, based on a generative adversarial network (GAN) and a convolutional neural network (CNN), to classify defects of a textured surface, while a novel label assignment scheme is proposed to integrate unlabeled samples into semi-supervised learning to enhance the overall performance of the system. In this framework, a customized GAN uses limited labeled samples to generate unlabeled ones, while the proposed label assignment scheme makes the generated data follow different label distributions in such a way that they can participate in training with labeled data. Finally, a CNN is proposed for semi-supervised training and the category identification of each defect sample. Experimental results show the effectiveness and robustness of the proposed framework even if original samples are limited. We verify our approach on four different surface defect datasets, achieving consistently competitive performances.

Keywords: deep learning (DL); generative adversarial network (GAN); label assignment; semi-supervised learning

Citation: He, Y.; Wen, X.; Xu, J. A Semi-Supervised Inspection Approach of Textured Surface Defects under Limited Labeled Samples. *Coatings* **2022**, *12*, 1707. https://doi.org/10.3390/coatings12111707

Academic Editors: Luca Luca Vattuone and Michael Nolan

Received: 16 September 2022
Accepted: 7 November 2022
Published: 9 November 2022

Publisher's Note: MDPI stays neutral with regard to jurisdictional claims in published maps and institutional affiliations.

Copyright: © 2022 by the authors. Licensee MDPI, Basel, Switzerland. This article is an open access article distributed under the terms and conditions of the Creative Commons Attribution (CC BY) license (https://creativecommons.org/licenses/by/4.0/).

1. Introduction

Defect classification is a fundamental industrial inspection task, the aim of which is to identify the category of a defective image. It is commonly performed on textured surfaces of many industrial products, such as metal [1], wood [2], and fabrics [3–5]. This process is crucial to guarantee product quality but is always executed manually in practice. Aiming at replacing this human-involved operation, there are many approaches using machine learning or deep learning techniques in automatic defect classification. However, these methods are mainly based on a supervised scenario, where all the data must be labeled. Unfortunately, the defect samples not only are hard to collect in large numbers but also need to be labeled by experts. Supervised approaches have no ability to handle unlabeled samples, which can lead to defect misclassification. This work introduces a semi-supervised learning (SSL) approach that can generate and handle unlabeled defect data in such a way that outperforms the supervised ones which only deal with labeled data. Therefore, the SSL approach can achieve reliable accuracy with limited samples, and the heavy and complicated manual sample collection and label allocation work can be alleviated.

To achieve automatic defect classification, the previous work on addressing this problem can be categorized into traditional shallow network learning (SNL)-based approaches and deep network learning (DNL)-based approaches. For simplicity, they are called SNL-based and DNL-based approaches, respectively.

SNL-based methods: In [6], Zhang et al. extracted the features based on Fisher distance and principal component analysis (PCA), and used the support vector machine (SVM) classifier for defect detection on an aluminum alloy surface. Yunwon and Kweon [7] proposed a neighboring difference filter algorithm to extract the foreground defective regions and used a random forest for defect classification. In [8], a guidance template-based algorithm using the statistical characteristic of textures is proposed for defect classification of strip steel surface. In [9], the combination of the completed local binary pattern (CLBP) features and the nearest-neighbor classifier are used to perform defect classification tasks. It is noteworthy that studies [7–9] are the combination of a hand-crafted feature extractor and a typical ML classifier, which highly rely on human experience.

DNL-based methods: Zhou et al. [10] used a simple sequential structured CNN for feature extraction and fed the representation tensors into a softmax layer for classification. Hossain et al. [11] proposed a light model of a six-layer CNN for fruit classification. Huang et al. [12] designed a small FCN network a quarter of the size of the original network [13]. These methods usually had to use tiny CNN structures because of limited industrial data. Except for the self-designed networks, some approaches adopted various baseline CNNs for defect classification via transfer learning. Through pre-training on the large dataset ImageNet, large baseline CNNs with dozens of layers, such as AlexNet [14], VGG [15], Inception [16], and ResNet [17], can be applied to specific tasks and avoid overfitting. In [11], the classification system used not only a tiny model but also fine-tuned a pre-trained VGG16 model. Studies [18,19] directly used pre-trained models for representation learning and then fine-tuned them by their own data for defect classification. The approach presented in [18] used the ZFNet to classify various textured surface defects and gave further evidence that ImageNet pre-trained models can be applied in industrial inspection tasks, whereas the distributions of textured defect data differ from those of ImageNet data. Yang et al. in [19] presented the transfer learning method for surface defect classification of flat panel displays. The study used AlexNet as the backbone and reduced one fully connected layer for online training. Using large models can achieve higher precision but also cost more computation sources because of many model parameters, especially in the last fully connected layers. Therefore, some approaches combined the DL and ML techniques, which regarded the convolutional part as a feature extractor, with the aim to replace hand-crafted features, and then are fed into a typical ML classifier. Studies [12,20] combined the transferred CNN features and various ML classifiers in pursuit of a more economical inspection system. Natarajan et al. in [20] used a combination of the VGG [15] model and the SVM classifier for metal surface defect classification, where VGG was employed to extract feature maps which were then used as input to SVM. In [12], hierarchies of features were extracted by FCN models and a two-stream algorithm was proposed to classify defects, which can only be applied on single-class image. The combination-like strategy aims to replace hand-crafted features with CNN features, which have a higher abstract level and stronger robustness [21]. However, this also resulted in poor generalization and an extra training process for the ML classifier. The main drawback of the previous approaches is that they highly depend on supervised learning, which requires enough defect samples and can only process labeled ones. The above approaches, whether based on fully supervised learning or transfer learning, all fail to provide correct classification in the presence of unlabeled samples.

Therefore, this study attempts to establish an SSL defect classification system, which can handle both labeled and unlabeled samples. Indeed, for an SSL scheme, the first consideration is the acquisition of unlabeled samples. As in other fields [22–24], the semi-supervised or weakly supervised learning methods can work on enough unlabeled samples that have become available on the Internet. However, it seems impossible to collect the same scale of defect data due to the rare occurrence of defects and the privatization of available data. Therefore, it is a better choice to consider how to generate valuable samples instead of taking much time to collect real ones. In this paper, a standard generative adversarial network (GAN) architecture is customized to fit defect data and then used to generate unlabeled samples [25]. Based on original defect samples, the GAN can generate new ones through a competitive training process involving a pair of networks.

When there are enough unlabeled samples, the second problem is how to include them in training. Different from transfer learning that trains labeled and unlabeled samples alternately, the SSL system needs to train both simultaneously. The unlabeled samples are often assigned to weak or pseudo labels for SSL, which is not a good choice for GAN samples. The unlabeled samples they used, such as in studies [23,26], are real but GAN samples are fake data generated by training. The existing methods of processing GAN samples are too rough, treating them the same way during training—the GAN samples are regarded as an extra class [27] or placed directly into existing classes [28].

We consider that all the GAN samples cannot be regarded as identical; the high-quality ones should be used to expand the original dataset and the low-quality ones can boost the learning process but not affect its optimization direction. This study proposes the label assignment for the unlabeled samples (LAUS) algorithm, which makes different assumptions for the unlabeled samples. The high-quality samples regarded as real images are assigned the corresponding ground-truth class distributions, and the low-quality ones are assigned a uniform label distribution over all the ground-truth classes, which assumes that these samples do not belong to a specific class. In this way, labeled and unlabeled samples can be mixed together for training. Moreover, since there are enough samples for training, the CNNs used in defect classification will not be subject to tiny networks [10–12] or large baseline networks [14,15,17], and hence can be designed more flexibly and scientifically. This study designed a new CNN used as the classifier, named all learning lightweight network (ALLnet), which focuses on industrial gray images, the layers of which are all learnable. Compared with the CNNs used in previous works, this designs a suitable trade-off between model power and size.

To address these two challenges, this paper introduces an SSL approach to classify defects of textured surfaces with very few labeled samples. This method uses a GAN for sample generation, a label assignment algorithm for semi-supervised training, and a CNN for representation learning. Unlike the previous supervised ones, our SSL method can generate unlabeled samples by itself and has the ability to handle labeled and unlabeled samples simultaneously, and thereby achieves higher accuracy and robustness. To verify our approach, we carried out extensive experiments on four different defect datasets. The main contributions of this work are summarized as follows:

(1) An SSL framework that integrates a customized GAN to generate new samples is proposed to classify defects under limited labeled samples.
(2) A label assignment algorithm that includes the unlabeled samples generated by GAN into training together with labeled ones is proposed.
(3) A detailed analysis on the CNNs used in defect inspection and the network ALLnet is designed for representation learning that is used in our SSL pipeline.

2. Unlabeled Samples Generation

2.1. Standard GAN

GANs are the recent emerging deep architectures for both semi-supervised and unsupervised learning [25]. Unlike other DL networks, the GAN learns around two subnetworks, a generator G and a discriminator D, and thus it can be characterized by training these two networks in competition with each other. In each training step, G produces a sample from a random noise z, with the aim of fooling the D. The D receives the generated samples as well as the real data x to classify them as "real" or "fake". Subsequently, G is devoted to producing more realistic images and D works to improve the "distinguishability". Both networks are updated repeatedly, and the iteration stops when they reach a Nash equilibrium. In more detail, D and G are competitors in a minimax game with the following function.

$$\min_G \max_D V(G,D) = E_{p_{data}(x)} \log D(x) + E_{p_z(z)} \log[1 - D(G(z))] \quad (1)$$

where E is the empirical estimate of the expected value of the probability. G transforms z into $G(z)$, which is sampled from a noise distribution p_z, and the ideal p_z should converge to the real data distribution p_{data}.

2.2. The Customized GAN

As already mentioned, the GAN has two sub-networks, a generator and a discriminator, which can be multi-layer perceptron [25], auto-encoders [29], or CNNs [30]. The customized GAN in our system uses CNNs as a backbone and is designed to be applicable to industrial gray images.

The GAN architecture is shown in Figure 1. For the generator, we feed into a 100 d random noise vector and reshape it to $4 \times 4 \times 8$ using a linear function (because the mini-batch size is 64, the actual input size is $4 \times 4 \times 512$ for each past). To enlarge the tensors, four deconvolutional layers are used with a kernel size of 5×5 and a stride of 2, denoted as {D1, D2, D3, D4}. Each deconvolutional layer follows a BN layer and a ReLU function, except the D4, which uses the tanh function. Finally, a sample that is $64 \times 64 \times 1$ in size can then be generated. The discriminator receives the generated samples and real images as input. Similar to the generator, we use four convolutional layers to classify whether the input image is real or fake. These layers are also 5×5 in size with a stride of 2, denoted as {C1, C2, C3, C4}, each of which is equipped with a BN layer and a LeakyReLU function. The settings of output activation functions are followed by the outstanding conclusions in [27]. We add two fully connected layers to receive the last convolution feature maps and then feed them into a sigmoid output.

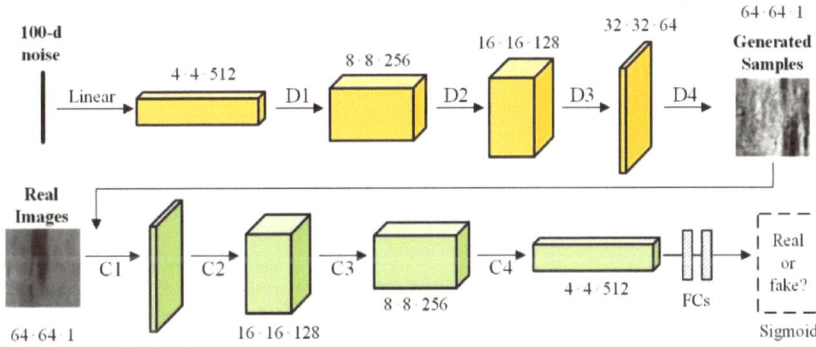

Figure 1. The architecture of the customized GAN.

For all the datasets, we train the GAN on the same hyper-parameters. We use stochastic gradient descent (SGD) to train the models in the GAN with a mini-batch size of 64. All the weights are initialized from a normal distribution; the variance is 0 and the standard deviation is 0.02. The slope of leak is set to 0.2 in the LeakyReLU function. For each dataset, the GAN will be trained with a learning rate of 0.0001 for 600 epochs.

3. Methodology

3.1. Overview of the SSL Framework

Figure 2 shows an overview of the proposed SSL framework, which consists of three parts: the customized GAN, the LAUS, and the ALLnet. In a single pass, the GAN receives labeled samples and noise as input and then generates unlabeled samples. Next, using the LAUS algorithm, it processes the generated samples in such a way that they can be included into labeled ones as training data. Finally, the ALLnet is trained on the mixed samples and thus has high defect classification ability.

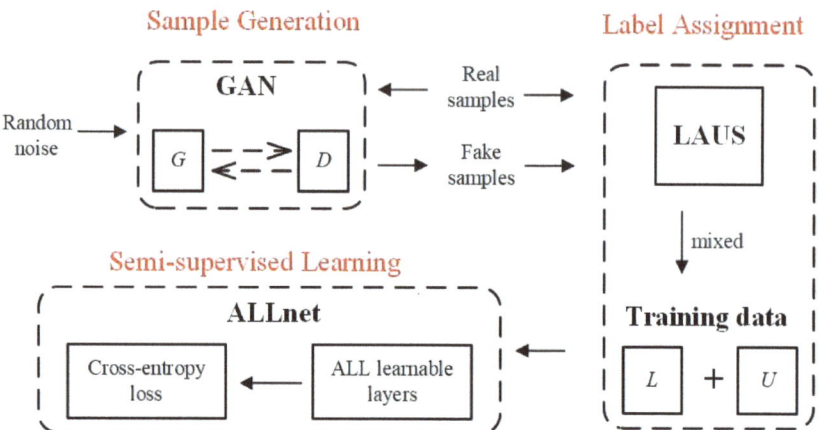

Figure 2. The proposed SSL framework. "L" and "U" represent labeled samples and unlabeled samples.

3.2. LAUS Algorithm

In order to train with labeled samples, an SSL system makes a reasonable assumption on unlabeled samples and assigns them "labels"—often not the real labels. There are two assumptions that are commonly used for unlabeled samples. One is the "another label", which creates an extra class label for the unlabeled samples [27]. The other is the "pseudo label", which assigns each generated sample a ground-truth label according to the prediction output [28]. Both methods are effective, but their assumptions are too rough. For the unlabeled samples generated by GAN, "another label" considers that they have poor quality and places them into a new class outside existing ones, whereas "pseudo label" considers their quality as good as real samples and as belonging to the existing classes. In truth, however, the image quality produced by GAN is unpredictable, and the ratio of high-quality samples to low-quality samples in each category is also different.

Instead of creating a new class or pseudo labels, the proposed LAUS aims to isolate the whole GAN sample and assign them to different label distributions. Through the model that was trained on original samples, each GAN sample can obtain the corresponding class probability vector, the maximum of which will be treated as its class score. The high-score samples are assigned to their corresponding ground-truth class distributions. These samples join the existing classes, act as real data, and extend the labeled samples. The low-score samples are assigned to a uniform label distribution over all the existing classes. These samples, therefore, belong to neither the existing classes nor a new class (see Figure 3). After the LAUS process, the GAN samples that share common knowledge with the real samples can be included in training and not affect the learning direction towards unknown or fake classes. With this strategy, we can train powerful models on enough defect data and avoid the risk of over-fitting.

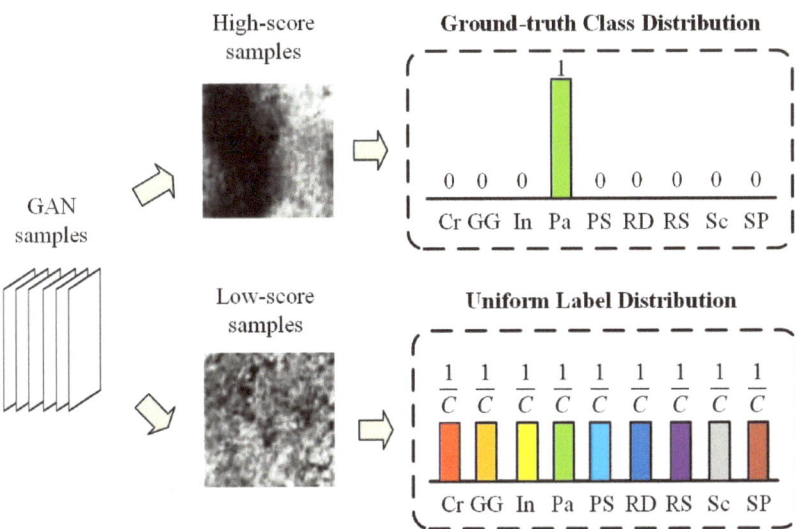

Figure 3. The label distributions for different generated samples.

First of all, a CNN is supervised trained (i.e., the ALLnet in our pipeline) on real samples S_r, and the output feature maps can be written as:

$$z_j = \sum_{i \in M_j} \sigma(w_i, b_i) \tag{2}$$

where w and b are the weights and bias of the i-th neuron in the j-th convolutional layer, and M_j is the set of input feature maps of the j-th convolutional layer. $\sigma(\cdot)$ represents the activation function. After the training is completed, this model Ms is used to obtain the class score of the GAN sample. The output of the ALLnet is the softmax function that is defined as:

$$\theta_j = \frac{e^{z_j}}{\sum_{k \in C} e^{z_k}} \tag{3}$$

For the GAN-generated samples (SGAN), the class score τ of each SGAN is the maximum prediction probability, $\tau = \text{argmax}\,(\theta_1, \ldots, \theta_j), j \in C$, where C is the number of ground-truth classes. According to τ, the SGAN are distributed into two sets of training data: labeled samples S_l and unlabeled samples S_u. The generated samples with high scores join S_l and receive the ground-truth class label "y", corresponding to the maximum score. They are assigned the ground-truth class distribution q_{GC} as well as the S_r, and the q_{GC} can be written as:

$$q_{GC} = \begin{cases} 0, k \neq y \\ 1, k = y \end{cases} \tag{4}$$

The rest of the generated samples are assigned the uniform label distribution q_{UL} that can be written as:

$$q_{UL} = \frac{1}{C} \tag{5}$$

Furthermore, if the real samples are so few that the start model M_s experiences serious over-fitting, the generated images would be regarded as unlabeled samples and assigned to the uniform label distribution. Finally, the LAUS transforms all the GAN samples into trainable data, and the details of it appear in Algorithm 1.

Algorithm 1: LAUS Algorithm

Input: GAN samples S_{GAN}, real samples S_r, and ground-truth class labels θ_{gt}.
1:    Train ALLnet on $\{S_r, \theta_{gt}\}$ => startup model M_S.
2:    **for** i **in** S_{GAN}:
3:        **if** no M_S:
4:            Label S_{GAN} as a uniform label distribution $\mathcal{L}_{uniform}$.
5:            $S_u \leftarrow S_{GAN}$, $S_l \leftarrow S_r$.
6:            **break**
7:        **end if**
8:        use M_S on $S_{GAN}^{(i)}$ => class scores $\tau^{(i)}$.
9:        **if** $\tau^{(i)} \geq \tau_{threshold}$:
10:           label $S_{GAN}^{(i)}$ as $\mathcal{L}_{gt}$ according to $\tau^{(i)}$.
11:           $S_u \leftarrow S_u + S_{GAN}^{(i)}$.
12:       **else:**
13:           label S_{GAN} as $\mathcal{L}_{uniform}$.
14:           $S_l \leftarrow S_l + S_{GAN}^{(i)}$.
15:       **end if**
16:       $S_l \leftarrow S_r$.
17:       Training data $S_T \leftarrow S_u + S_l$.
18:   **end for**
19:   **Return** S_T

3.3. ALLnet

The baseline CNNs have become the common solution to deal with the representation learning on big data. These heavyweight networks, although they can achieve adequate results, may be unsuitable for some industrial scenes. For industrial inspection tasks, there are not many categories in one defect dataset in general, and therefore, we neither have enough defect data to train large networks nor need a network with so many neurons. In this context, we propose a novel network, ALLnet, as the representation learning machine of the semi-supervised pipeline. There are four blocks in ALLnet, denoted as {B1, B2, B3, B4}. Each block consists of three learnable layers: a 3 × 3 conv with a stride of 1, a 3 × 3 conv with a stride of 2, and a BN layer. We stack the layers with small 3 × 3 conv kernels, which can enhance the model capacity and complexity in fewer parameters [15]. The spatial pooling layer is replaced with stride convolution that lets the network learn how to down-sample, and the fully connected (fc) layers on top of convolutional features are replaced with a global average pooling (GAP) layer which has nearly zero parameters.

By the LAUS, we mixed the labeled and unlabeled samples into training data. Then, they are fed into the ALLnet for SSL until the maximum iteration is reached.

3.4. Training
3.4.1. Loss Function

In this paper, we use the cross-entropy loss for the SSL system. Let $k \in \{1, 2, \ldots, C\}$ be the predicted class, where C is the number of ground-truth classes. The cross-entropy loss can be formulated as:

$$L = -\sum_{k=1}^{C} \log(p(k)) q(k) \qquad (6)$$

where $p(k)$ is the prediction probability that an input sample belongs to class k. It is derived from the softmax function, which normalizes the output of the previous fully connected layer. $q(k)$ is the ground-truth class distribution. Let y be the corresponding ground-truth class, and $q(k)$ can be defined as:

$$q(k) = \begin{cases} 0 & k \neq y \\ 1 & k = y \end{cases} \qquad (7)$$

Therefore, if we only consider the non-zero term, (6) can be equivalent to:

$$L = -\log(p(y)) \tag{8}$$

As in Algorithm 1, the GAN samples have different assumptions. The high-score samples are assigned to ground-truth class distribution, whereas the low-score samples are assigned to a uniform label distribution $q_{LAUS}(k)$, which is uniformed over the ground-truth classes. Let τ be the threshold of class score, and $q_{LAUS}(k)$ can be defined as:

$$q_{LAUS}(k) = \begin{cases} q(k) & p(k) \geq \tau \\ \frac{1}{C} & p(k) < \tau \end{cases} \tag{9}$$

where the $q(k)$ in the upper part is denoted in (7).

For the real images and the high-score samples, the cross-entropy loss is equivalent to (8), but for low-score samples, (6) can be equivalent to:

$$L = -\frac{1}{C} \sum_{k=1}^{C} \log(p(k)) \tag{10}$$

Combining (8), (10), and (6), the cross-entropy loss for our SSL framework can be rewritten as:

$$L = -(1-w)\log(p(y)) - \alpha(n)\frac{w}{C}\sum_{k=1}^{C}\log(p(k)) \tag{11}$$

If the input image is labeled, $w = 0$. If the input image is unlabeled, $w = 1$. Since there are more labeled samples than unlabeled ones in training data, a penalty function $\alpha(n)$ is incorporated into the unlabeled item, which can avoid making the learning tend to rapid deterioration when too many GAN samples are added. $\alpha(n)$ can be written as:

$$\alpha(n) = \begin{cases} 0 & n < N_1 \\ \frac{n-N_1}{N_2-N_1}\alpha_t & N_1 \leq n < N_2 \\ \alpha_t & N_2 \leq n \end{cases} \tag{12}$$

where n is the number of generated samples and αt is the threshold, which is set to 0.8 in this paper. For SSL, the number of unsupervised samples should not be less than that of the supervised ones. Let N_{real} be the number of real training images, N_1 is equal to N_{real}, and N_2 is five times the value of N_{real}.

3.4.2. Implementation

We train the SSL model to minimize the loss function in (11). All the input images, consisting of real and GAN ones, are resized as 64 × 64 × 1 and randomly mixed in training data. The mini-batch size of each input is 128. We train the models with the Adam optimizer [31] with the exponential decay parameters β_1 and β_2 set to 0.9 and 0.99, respectively. We train the model with a learning rate of 0.0001 for 100,000 mini-batch iterations. All the experiments were run in Python with Tensorflow packages on an Intel Core i7, 3.3 GHz with 64-GB RAM and a NVIDIA TITAN Xp GPU workstation.

4. Experiments

4.1. Defect Datasets

The proposed method is evaluated on four types of defect datasets, namely the wood defect dataset KNOTS [2], the textile dataset Fabrics [3], the steel plate defect dataset NEU-CLS-64 [32], and the magnetic tile defect dataset MT [33]. NEU-CLS-64 is improved on the defect dataset NEU-CLS, which contains 1800 images of six defect classes. The motive to recreate the NEU-CLS-64 is that multiple and even unknown defects exist in an image. So, we extended the NEU-CLS in terms of quantity and category. In more detail, every image was resized to 192 × 192 and center-cropped into nine 64 × 64 images. Then, we discarded the following cropped images: the ones with heavily occluded defects, edge-truncated defects, and no defect. Finally, the updated dataset NEU-CLS-64 assembles approximately 7000 tiny images with nine defect classes, i.e., crazing (Cr), grooves and gouges (GG), inclusion (In), patches (Pa), pitted surface (PS), rolling dust (RD), rolled-in scale (RS), scratches (Sc), and spots (Sp). The details of the NEU-CLS-64 and three other defect datasets are summarized in Table 1, and the examples of each dataset are shown in Figure 4. In this section, we mainly report results on the NEU-CLS-64, which is a relatively large-scale dataset; the other three datasets serve as auxiliary datasets for further verification. All the datasets use the same train/test ratio in experiments, and the ratio is 7:3 in this study.

Table 1. Details of four defect datasets.

Datasets (Num.)	Defect Type and Num.
NEU-CLS-64 (7226)	Cr (1210), GG (296), In (775), Pa (1148), PS (797), RD (200), RS (1589), Sc (773), SP (438).
KNOTS (425)	Dry (69), Edge (65), Encased (30), Horn (35), Leaf (47), Sound (179).
Fabrics (1173)	Cotton (588), Denim (162), Nylon (57), Polyester (226), Silk (50), Wool (90).
MT (392)	Blowhole (115), Break (85), Crack (57), Fray (32), Uneven (103).

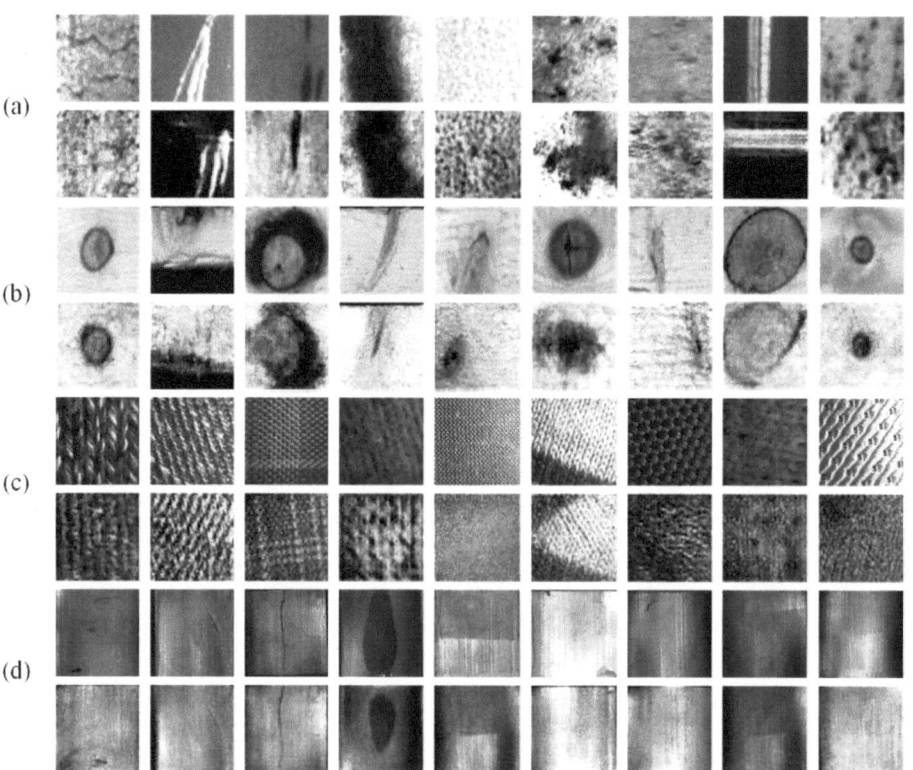

Figure 4. Examples of the real images and the GAN samples of the four defect datasets. In each group, the top row shows real images and the bottom row shows GAN samples. (**a**) NEU-CLS-64, (**b**) KNOTS, (**c**) Fabrics, (**d**) MT.

4.2. GAN Results

We first need to ensure that most samples generated by GAN are available. The examples of GAN samples are shown in Figure 4. For most types of defects, GAN samples cannot be intuitively distinguished from real images by human eyes. However, several images of some defect categories are fuzzy, such as the "crazing" in the NEU-CLS-64 (see the first column in Figure 4a). Next, we want to prove that the unlabeled GAN samples can boost the classification performance. Since unlabeled samples are relatively easy to obtain, we hope that these generated samples have similar semantic information to the real samples, and thereby can assist or replace the labeled samples that are too difficult to collect. In order to verify this, we added different numbers of GAN samples generated on the NEU-CLS-64 into training, and the results are given in Table 2. From this table, we can safely conclude that the unlabeled GAN samples can share common knowledge with the labeled ones, and by the addition of GAN samples, there is a significant increase in classification accuracy. As we discussed in Section 3, LAUS makes a more reasonable assumption for GAN samples and therefore achieves the best results. The classification accuracy is improved by approximately three points when $1 \times$ GAN samples are added compared to the original samples. As the number of GAN samples increases, classification accuracy increases gradually, and it reaches the peak when $3 \times$ GAN samples are added. Since the low-score samples are much more abundant than the high-score ones, the accuracy would tend to deteriorate if too many GAN samples are added, but even so, adding $5 \times$ GAN samples still gives a 3.73% improvement in accuracy over the baseline.

Table 2. Comparing LAUS with related algorithms.

$S_{GAN}:S_{real}$	LAUS	Another Label	Pseudo Label
	acc. (%)	acc. (%)	acc. (%)
0 (baseline)	96.24	96.24	96.24
$1\times$	98.04	98.19	98.39
$2\times$	98.40	98.60	98.55
$3\times$	99.37	98.92	98.41
$4\times$	99.13	99.04	97.90
$5\times$	98.97	99.02	97.61

We evaluated the GAN samples for each category of the NEU-CLS-64 in detail, and the results are shown in Figure 5. We wanted to explore how unlabeled samples contribute to overall accuracy when they are trained along with labeled samples. From Figure 5, we observe that the addition of GAN samples can lead to a consistent decrease in error rate for all the defect classes. Unsurprisingly, the ratio of high-score and low-score samples produced on each class is quite different. The class with a higher error rate seems to produce fewer high-score samples, but enough low-score samples can still reduce the error rate. Like the Cr, although it only has only 2% high-score samples of the GAN samples, the Top-1 error rate of Cr dropped dramatically by approximately 5%.

4.3. Comparison with Other Label Assignment Algorithms

We compare the LAUS with the "another label" and "pseudo label", which can also be used in the semi-supervised framework. The experimental results on the NEU-CLS-64 are also listed in Table 2. In this table, we observe that both methods are effective and the LAUS can also exceed them by 1~2 points when they all reach the peak. We consider the reason for this is that LAUS's assumption for GAN samples is more reasonable. The "another label" makes a coarse assumption for the GAN samples that are all considered as low-score ones. So, as the number of GAN samples increases, the accuracy rises gradually. The "pseudo label" gives an over-optimistic evaluation of the GAN samples and puts them all in the ground-truth classes. It causes the accuracy to peak before many GAN samples are added, and then drops sharply due to too many low-score samples making the learning process deteriorate. Unlike the onefold assumptions of the "another label" and "pseudo label", the LAUS performs different assignments for the GAN samples, which may be the reason why the LAUS has a superior performance.

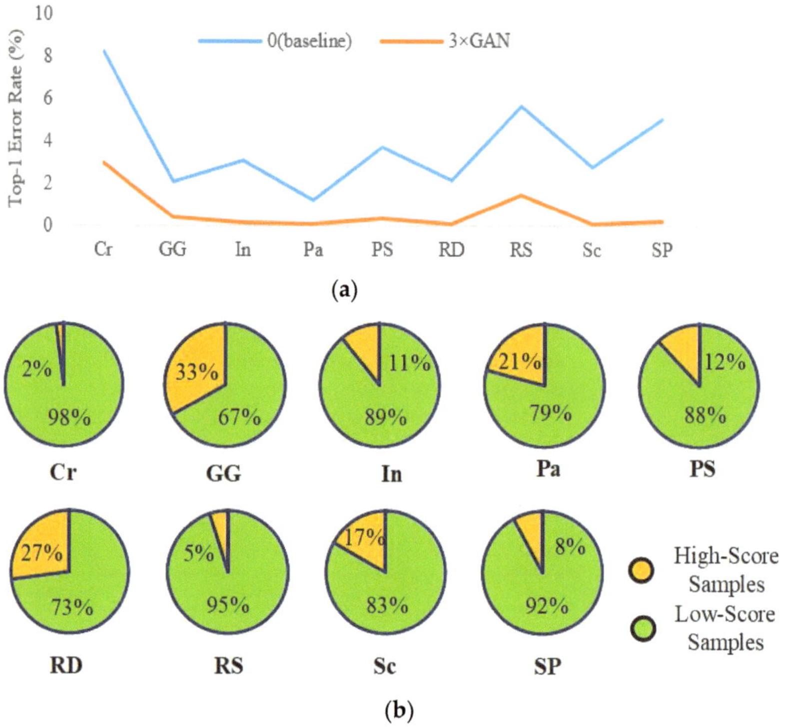

Figure 5. The effects of adding GAN samples on each defect class of NEU-CLS-64. (**a**) The Top-1 error rate comparison between no GAN samples added (baseline) and adding 3 × GAN samples. (**b**) The percentage of high-score samples and low-score samples of each defect class in the 3 × GAN samples.

4.4. Comparison with Baseline CNNs

We replace the ALLnet with other baseline CNNs to evaluate its effect in a semi-supervised framework. Since baseline CNNs require a great amount of training data, we enlarged the training set for a fair comparison, where 3 × GAN samples are added. The input size and weight parameters of CNNs used in experiments are different, and hence we use different training epochs and minibatch sizes for them. The time of a float operation is used as the measure of runtime. The AlexNet and VGG16 are trained with a minibatch size of 64 for 300 epochs, and the InceptionV3 and ResNet101 with a minibatch size of 32 for 600 epochs. The training configuration for ALLnet is defined in Section 3. The results on NEU-CLS-64 are listed in Table 3. It is no surprise that the strong baseline CNNs obtained adequate results—the accuracy exceeds 99%. However, a large input size requires enough memory and redundant parameters and has excessive computation costs, which can slow down the convergence and increase the computation time. By comparison, the ALLnet not only achieved the same level of accuracy as the large networks, but also has a simpler structure and less computing time. Furthermore, we also make a detailed assessment of the internal structure of the ALLnet, mainly to judge the impact of the number of blocks and input size on the model. We observe that each additional block in the ALLnet brings an approximately 1% increase in the overall accuracy, and this trend will continue until there are four blocks in the model. Another finding is that expanding input size can hardly increase accuracy but can greatly reduce computational speed. According to the above results, we can conclude that the ALLnet with four blocks achieved the best trade-off between speed and precision, and a large network seems to be unnecessary for industrial inspection tasks.

Table 3. Comparison with baseline CNNs.

Models	Input Size	Parameters	Accuracy (%)	Runtime (ms/Float)
AlexNet	224 × 224	5.8×10^7	99.10	42.1
VGG16	224 × 224	13×10^7	99.62	66.2
InceptionV3	299 × 299	2.4×10^7	99.29	21.0
ResNet101	224 × 224	4.0×10^7	99.47	29.7
ALLnet(2B)	64 × 64	0.2×10^6	97.50	6.9
ALLnet(3B)	64 × 64	0.7×10^6	98.44	8.4
ALLnet(4B) *	64 × 64	2.9×10^6	99.37	8.6
ALLnet(4B)	128 × 128	2.9×10^6	99.37	21.0
ALLnet(5B)	64 × 64	12×10^6	99.38	10.2

* represents that this model is used in our semi-supervised framework. "B" indicates the block of the ALLnet, which consists of two conv layers and a BN layer.

4.5. Comparison with Other Defect Classifiers

We compare our SSL method with state-of-the-art defect classifiers based on supervised learning or transfer learning. The methods selected for comparison include the supervised ones by Zhou et al. [12] and Hossain et al. [13], and the transfer learning-based ones are Decaf [14] and MVM-VGG [16]. For simplicity, we call these four methods S1, S2, T1, and T2, respectively. These methods are reproduced by Tensorflow as our method for fair comparison. The contrast experiments were carried out on the four aforementioned defect datasets by adding different numbers of GAN samples. These GAN samples were added in two ways: one is to compensate the difference between classes to restore a dataset into a balanced one (balance mode, "B" for short); the other is to follow the original inter-class ratio (unbalance mode, "U" for short). The results are shown in Table 4.

Table 4. Comparison with other defect classifiers on four defect datasets.

Dataset		NEU-CLS-64					KNOTS				
Methods		S1	S2	T1	T2	Ours	S1	S2	T1	T2	Ours
0 (baseline)		87.50	88.14	92.87	97.00	96.24	79.69	73.44	80.21	73.81	80.38
1 × GANs	B	93.25	92.04	96.84	97.87	98.04	84.38	89.06	85.28	89.31	89.06
	U	93.64	91.16	95.66	96.87	97.86	78.91	84.38	82.09	85.29	86.72
2 × GANs	B	95.62	95.40	97.76	98.58	98.40	90.63	89.84	88.69	92.30	91.14
	U	94.72	93.25	95.89	98.00	98.79	87.50	85.90	83.56	88.69	90.63
3 × GANs	B	97.94	97.11	99.10	99.34	99.37	94.53	94.01	96.62	97.49	97.49
	U	96.63	95.16	97.60	98.89	98.87	93.49	91.41	92.20	94.62	95.31
4 × GANs	B	97.26	95.81	98.55	99.12	99.13	92.97	92.96	93.02	97.00	95.00
	U	96.13	94.28	96.16	92.87	98.36	88.80	90.89	90.55	92.37	94.53
5 × GANs	B	96.65	95.66	97.75	98.13	98.87	91.41	93.49	92.37	96.53	94.68
	U	84.90	-	91.34	90.43	84.91	88.67	91.46	86.32	93.02	83.33
Datasets		Fabrics					MT				
0 (baseline)		-	-	73.47	82.33	65.42	96.88	90.31	92.60	59.32	92.19
1 × GANs	B	67.71	73.70	83.33	90.55	80.02	91.41	92.19	94.88	66.87	96.09
	U	69.27	-	79.93	88.22	77.60	89.84	91.41	92.90	64.62	95.31
2 × GANs	B	86.91	85.32	90.10	91.70	88.93	92.58	-	97.51	79.62	97.40
	U	79.84	82.34	86.40	91.09	84.21	94.53	96.88	92.33	64.99	92.19
3 × GANs	B	90.94	91.99	91.90	93.07	93.80	95.31	-	99.09	82.84	99.22
	U	90.00	90.04	89.01	92.80	90.31	99.22	98.18	96.68	81.63	97.65
4 × GANs	B	87.30	89.55	91.04	91.61	91.40	97.92	-	98.82	82.77	98.17
	U	88.09	86.33	88.64	91.54	90.09	97.66	97.66	95.31	82.59	97.39
5 × GANs	B	85.68	89.50	87.99	92.98	90.04	96.61	-	97.79	83.82	98.88
	U	83.33	82.50	87.50	91.15	86.98	97.40	97.27	95.01	85.77	94.27

In the four defect datasets, our method achieves 3.13%, 17.11%, 28.38%, and 7.03% improvements in accuracy compared to the baseline, respectively. Meanwhile, our method at the point of 3 × GANs consistently obtained the best results and is superior to other methods for each dataset. For each dataset, our method can work well even if the original data are limited, which also shows that our method performs better than other methods in terms of generalization and robustness. Moreover, our method can easily correct the original unbalanced defect dataset into a balanced one, which brings an improvement in accuracy.

5. Conclusions

In this paper, we propose a semi-supervised learning method that mainly deals with data-limited defect classification tasks. This method has no need for extra collection of defect data but uses a customized GAN to generate samples. Through the proposed LAUS, the GAN samples can be trained with the limited original samples simultaneously. We also designed the ALLnet, which is trained on these samples in a semi-supervised manner. Extensive experiments on four different defect datasets have shown that under the semi-supervised learning framework, we obtained substantial accuracy improvements that range from 3.13% to 28.38%. Our method can be more precise and robust than the previous state-of-the-art transfer learning and supervised learning methods, and is effective for defect classification when the original samples are limited.

Author Contributions: Conceptualization, Y.H.; Data curation, J.X.; Funding acquisition, Y.H.; Resources, X.W. All authors have read and agreed to the published version of the manuscript.

Funding: This research was funded by the SUT Youth Scientific Research Ability Cultivation Project, grant number 200005796.

Institutional Review Board Statement: Not applicable.

Informed Consent Statement: Not applicable.

Data Availability Statement: Not applicable.

Conflicts of Interest: The authors declare no conflict of interest.

References

1. Yu, H.; Li, Q.; Tan, Y.; Gan, J.; Wang, J.; Geng, Y.-A.; Jia, L. A Coarse-to-Fine Model for Rail Surface Defect Detection. *IEEE Trans. Instrum. Meas.* **2018**, *68*, 656–666. [CrossRef]
2. Niskanen, M.; Kauppinen, H. Wood inspection with non-supervised clustering. *Mach. Vis. Appl.* **2003**, *13*, 275–285. [CrossRef]
3. Kampouris, C.; Zafeiriou, S.; Ghosh, A.; Malassiotis, S. Fine-grained material classification using micro-geometry and reflectance. In Proceedings of the European Conference on Computer Vision (ECCV), Amsterdam, The Netherlands, 8–12 October 2016; pp. 778–792.
4. Kahraman, Y.; Durmuşoğlu, A. Deep learning-based fabric defect detection: A review. *Text. Res. J.* **2022**. [CrossRef]
5. Kahraman, Y.; Durmuşoğlu, A. Classification of Defective Fabrics Using Capsule Networks. *Appl. Sci.* **2022**, *12*, 5285. [CrossRef]
6. Zhang, Z.; Wen, G.; Chen, S. Audible Sound-Based Intelligent Evaluation for Aluminum Alloy in Robotic Pulsed GTAW Mechanism, Feature Selection, and Defect Detection. *IEEE Trans Ind. Inf.* **2018**, *14*, 2973–2983. [CrossRef]
7. Park, Y.; Kweon, I.S. Ambiguous Surface Defect Image Classification of AMOLED Displays in Smartphones. *IEEE Trans. Ind. Inform.* **2016**, *12*, 597–607. [CrossRef]
8. Wang, H.Y.; Zhang, J.; Tian, Y.; Chen, H.Y.; Sun, H.X.; Liu, K. A Simple Guidance Template-Based Defect Detection Method for Strip Steel Surfaces. *IEEE Trans. Ind. Inform.* **2018**, *15*, 2798–2809. [CrossRef]
9. Luo, Q.; Sun, Y.; Li, P.; Sun, Y.; Li, P.; Simpson, O.; Tian, L.; He, Y. Generalized Completed Local Binary Patterns for Time-Efficient Steel Surface Defect Classi-fication. *IEEE Instrum. Meas* **2018**, *68*, 667–679. [CrossRef]
10. Zhou, S.; Chen, Y.; Zhang, D.; Xie, J.; Zhou, Y. Classification of surface defects on steel sheet using convolutional neural networks. *Mater. Teh.* **2017**, *51*, 123–131. [CrossRef]
11. Hossain, M.S.; Al-Hammadi, M.; Muhammad, G. Automatic Fruit Classification Using Deep Learning for Industrial Ap-plications. *IEEE Trans. Ind. Inf.* **2019**, *15*, 1027–1034. [CrossRef]
12. Huang, H.-W.; Li, Q.-T.; Zhang, D.-M. Deep learning based image recognition for crack and leakage defects of metro shield tunnel. *Tunn. Undergr. Space Technol.* **2018**, *77*, 166–176. [CrossRef]
13. Wen, W.; Xia, A. Verifying edges for visual inspection purposes. *Pattern Recognit. Lett.* **1999**, *20*, 315–328. [CrossRef]
14. Krizhevsky, A.; Sutskever, I.; Hinton, G.E. ImageNet classification with deep convolutional neural networks. In Proceedings of the ProceNeural Information Processing Systems (NIPS), Lake Tahoe, NV, USA, 3–6 December 2012; pp. 1097–1105.

15. Simonyan, K.; Zisserman, A. Very deep convolutional networks for large-scale image recognition. In Proceedings of the International Conference on Learning Representations (ICLR), San Diego, CA, USA, 7–9 May 2015.
16. Szegedy, C.; Vanhoucke, V.; Ioffe, S.; Shlens, J.; Wojna, Z. Rethinking the Inception Architecture for Computer Vision. In Proceedings of the IEEE Conference on Computer Vision and Pattern Recognition (CVPR), Las Vegas, NV, USA, 27–30 June 2016; pp. 2818–2826. [CrossRef]
17. He, K.; Zhang, X.; Ren, S.; Sun, J. Deep residual learning for image recognition. In Proceedings of the 2016 IEEE Conference on Computer Vision and Pattern Recognition (CVPR), Las Vegas, NV, USA, 27–30 June 2016; pp. 770–778. [CrossRef]
18. Ren, R.; Hung, T.; Tan, K.C. A Generic Deep-Learning-Based Approach for Automated Surface Inspection. *IEEE Trans. Cybern.* **2017**, *48*, 929–940. [CrossRef] [PubMed]
19. Yang, H.; Mei, S.; Song, K.; Tao, B.; Yin, Z. Transfer-Learning-Based Online Mura Defect Classification. *IEEE Trans. Semicond. Manuf.* **2017**, *31*, 116–123. [CrossRef]
20. Natarajan, V.; Hung, T.-Y.; Vaikundam, S.; Chia, L.-T. Convolutional networks for voting-based anomaly classification in metal surface inspection. In Proceedings of the IEEE International Conference on Industrial Technology (ICIT), Toronto, ON, Canada, 22–25 March 2017; pp. 986–991. [CrossRef]
21. Lecun, Y.; Bengio, Y.; Hinton, G. Deep learning. *Nature* **2015**, *521*, 436–444. [CrossRef]
22. Liu, C.; Su, K.; Yang, L.; Li, J.; Guo, J. Detection of Complex Features of Car Body-in-White under Limited Number of Samples Using Self-Supervised Learning. *Coatings* **2022**, *12*, 614. [CrossRef]
23. Papandreou, G.; Chen, L.-C.; Murphy, K.P.; Yuille, A.L. Weakly-and semi-supervised learning of a deep convolutional network for semantic image segmentation. In Proceedings of the International Conference on Computer Vision. (ICCV), Santiago, Chile, 13–16 December 2015; pp. 1742–1750.
24. Song, K.; Wang, J.; Bao, Y.; Huang, L.; Yan, Y. A Novel Visible-Depth-Thermal Image Dataset of Salient Object Detection for Robotic Visual Perception. *IEEE/ASME Trans. Mechatron.* **2022**, 1–12. [CrossRef]
25. Wu, H.; Prasad, S. Semi-Supervised Deep Learning Using Pseudo Labels for Hyperspectral Image Classification. *IEEE Trans. Image Process* **2017**, *27*, 1259–1270. [CrossRef]
26. Odena, A. Semi-Supervised Learning with Generative Adversarial Networks. In Proceedings of the International Conference on Machine Learning. (ICML), New York, NY, USA, 6–11 June 2015.
27. He, D.; Xu, K.; Zhou, P.; Dongdong, Z. Surface defect classification of steels with a new semi-supervised learning method. *Opt. Lasers Eng.* **2019**, *117*, 40–48.
28. Larsen, A.B.L.; Kaae, S.S.; Winther, O. Autoencoding beyond pixels using a learned similarity metric. In Proceedings of the International Conference on Machine Learning. (ICML), New York, NY, USA, 19–24 June 2016; pp. 1558–1566.
29. Alec, R.; Luke, M.; Soumith, C. Unsupervised representation learning with deep convolutional generative adversarial networks. *arXiv* **2015**, arXiv:1511.06434.
30. Li, P.; Chen, Z.; Yang, L.T.; Zhang, Q.; Deen, M.J. Deep Convolutional Computation Model for Feature Learning on Big Data in Internet of Things. *IEEE Trans. Ind. Inform.* **2017**, *14*, 790–798. [CrossRef]
31. Kingma, P.D.; Ba, J. Adam: A Method for Stochastic Optimization. In Proceedings of the International Conference on Learning Representations (ICLR), San Diego, CA, USA, 7–9 May 2015.
32. Song, K.; Yan, Y. A noise robust method based on completed local binary patterns for hot-rolled steel strip surface defects. *Appl. Surf. Sci.* **2013**, *285*, 858–864. [CrossRef]
33. Huang, Y.; Qiu, C.; Guo, Y. Saliency of magnetic tile surface defects. In Proceedings of the 14th IEEE International Conference on Auto-Mation and Engineering, Munich, Germany, 18–22 June 2018.

Article

TSSTNet: A Two-Stream Swin Transformer Network for Salient Object Detection of No-Service Rail Surface Defects

Chi Wan [1,2,3], Shuai Ma [1,2,3] and Kechen Song [1,2,3,*]

1. School of Mechanical Engineering & Automation, Northeastern University, Shenyang 110819, China
2. National Frontiers Science Center for Industrial Intelligence and Systems Optimization, Northeastern University, Shenyang 110819, China
3. Key Laboratory of Data Analytics and Optimization for Smart Industry, Ministry of Education, Shenyang 110819, China
* Correspondence: songkc@me.neu.edu.cn

Abstract: The detection of no-service rail surface defects is important in the rail manufacturing process. Detection of defects can prevent significant financial losses. However, the texture and form of the defects are often very similar to the background, which makes them difficult for the human eye to distinguish. How to accurately identify rail surface defects thus poses a challenge. We introduce salient object detection through machine vision to deal with this challenge. Salient object detection locates the most "significant" areas of an image using algorithms, which constitute an integral part of machine vision inspection. However, existing saliency detection networks suffer from inaccurate positioning, poor contouring, and incomplete detection. Therefore, we propose an innovative deep learning network named Two-Stream Swin Transformer Network (TSSTNet) for salient detection of no-service rail surface defects. Specifically, we propose a two-stream encoder—one stream for feature extraction and the other for edge extraction. TSSTNet also includes a three-stream decoder, consisting of a saliency stream, edge stream, and fusion stream. For the problem of incomplete detection, we innovatively introduce the Swin Transformer to model global information. For the problem of unclear contours, we expect to deepen the understanding of the difference in depth between the foreground and background through the learning of contour maps, so the contour alignment module (CAM) is created to deal with this problem. Moreover, to make the most of multimodal information, we suggest a multi-feature fusion module (MFFM). Finally, we conducted comparative experiments with 10 state-of-the-art (SOTA) approaches on the NRSD-MN datasets, and our model performed more competitively than others on five metrics.

Keywords: no-service rail surface defect; salient object detection; two-stream encoder; transformer; contour information

Citation: Wan, C.; Ma, S.; Song, K. TSSTNet: A Two-Stream Swin Transformer Network for Salient Object Detection of No-Service Rail Surface Defects. *Coatings* **2022**, *12*, 1730. https://doi.org/10.3390/coatings12111730

Academic Editor: Ajay Vikram Singh

Received: 19 October 2022
Accepted: 8 November 2022
Published: 12 November 2022

Publisher's Note: MDPI stays neutral with regard to jurisdictional claims in published maps and institutional affiliations.

Copyright: © 2022 by the authors. Licensee MDPI, Basel, Switzerland. This article is an open access article distributed under the terms and conditions of the Creative Commons Attribution (CC BY) license (https://creativecommons.org/licenses/by/4.0/).

1. Introduction

Rail quality inspection is very important in the rail production process in steel mills, and one of the most critical aspects is the detection of rail surface defects. Earlier detection of no-service rail surface defects can prevent economic losses and safety accidents from occurring in time.

The rails to be inspected are divided into the in-service and no-service rails, which usually have different defect maps. Images of in-service rail defects often have bright backgrounds, prominent weaknesses, and distinct contours. However, no-service rail defect maps often have dark backgrounds, uneven lighting, and impurities of various origins interfering with identification, which can easily cause different shapes of rail surface defects during the processing:

(1) When the rail is heated, due to unreasonable technology and techniques, thermal stresses are created in the rail material, forming cracks on the surface;

(2) If the billet is not cleaned or is cleaned improperly, the impurities attached to the billet remain on the surface of the finished rail after heating and rolling deformation. This is known as scarring;
(3) The presence of linear or curved grooves of varying depths on the surface of the rails, either continuously or intermittently distributed on the local surface, is known as a scratch. This is usually caused by improper installation of equipment during the rolling process.

The cracks, scarring, and scratches have different contours and defect depths, making the detection of no-service rail surface defects much more difficult than detecting in-service rail surface defects. Specific comparison pictures [1,2] are shown in Figure 1, where defects are marked with a red dotted line.

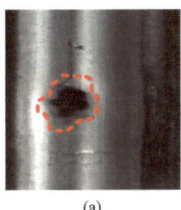 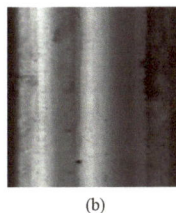 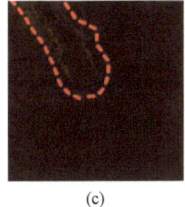 

(a) (b) (c) (d)

Figure 1. (**a**) Surface view of an in-service rail with defects [1]. (**b**) Surface view of an in-service rail [1]. (**c**) Surface view of a no-service rail with defects [2]. (**d**) Surface view of a no-service rail [2].

Today, more and more researchers from different fields are willing to work in this area, and they are proposing various solutions. Xu et al. [3] came up with a multi-frequency electromagnetic system to detect surface defects with different characteristics using electromagnetic waves of different frequencies. Hao et al. [4] proposed a new adaptive Canny algorithm without manually setting the parameters. Cao et al. [5] built a machine vision detection system using an improved least-squares method. Hao et al. [6] proposed a method to enhance the signal using improved Shannon entropy to reduce the noise generated by the track at high speeds. However, these methods can only detect superficial features, such as the color and texture of the defect.

As research into machine vision progresses, neural networks are becoming mainstream in the field of image classification and identification. Convolutional neural networks (CNNs) learn relationships between pixels using a series of operations such as convolution, pooling, etc., and can perceive differences in the depth gradient between the foreground and background of the image. As a result, research on CNN-based neural networks has become a hot topic. Shakeel et al. [7] proposed an adaptive multiscale attention module to align the feature maps. Zhang et al. [2] proposed MCNet using pyramid pooling to focus on a variety of contextual information. Baffour et al. [8] proposed a self-attention module working on the spatial locations. Among the many detection tasks, salient object detection (SOD) [9] locates the most noteworthy areas of a picture using vision algorithms such as human visual attention. SOD is important because it is often used as the first step of other vision tasks to focus on the most useful information in an image. In the current SOD deep learning networks, CNNs are often designed as the backbone of the network to abstract characteristics hierarchically. They tend to perform well on natural scene datasets because objects in natural scenes often have distinct contours, allowing the network to distinguish clearly between the foreground and background. At the same time, the network does not need to focus on long-range information, which also corresponds to the characteristics of CNNs' locality. However, in the field of rail surface defect detection, complex textures, the irregular outlines of defects, along with their blurred and dark edges, make CNNs' backbone produce incomplete defect recognition. Thus, we attempted to introduce a more competitive backbone into the field of rail defects detection—that is, the transformer [10].

In the past few years, the proposal of ViT [10] has made transformer a topic of interest in computer vision. ViT is famous for its ability to model global long-range dependency features. It uses a pure attention mechanism of computation, which reduces training time considerably compared to CNNs. Swin Transformer [11] has a CNN-like hierarchical feature structure comparable to that of ViT [10], and it calculates self-attention in a non-overlapped local window. It suggests connections between different local windows through an ingenious shifted window design. Swin Transformer [11] absorbs the locality, translation invariance, and hierarchical merits of CNNs, allowing it to be competent in visual tasks.

In view of these advantages of Swin Transformer [11], we introduced it to our network as the backbone. However, this does not solve all of the problems. The foreground and background textures of the rail surface are very similar, making it impossible for a single Swin Transformer [11] to locate defects accurately, and the resulting inspection map often has blurred edges. The experimental results are displayed in Figure 2.

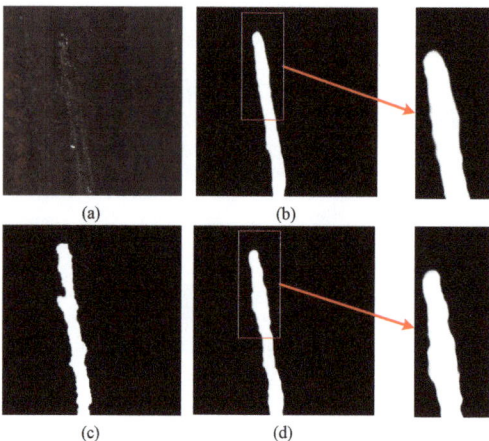

Figure 2. (**a**) Rail defect map. (**b**) Single Swin Transformer saliency map. (**c**) Ground truth (**d**) TSSTNet saliency map.

To deal with this problem, the existing network uses the first few stages for auxiliary tasks, such as boundary detection, and the last few for the main task, i.e., saliency detection [12–15]. However, in the experiments, we found that when multitasking with traditional single-stream encoders, the auxiliary tasks interfere with the primary task. Meanwhile, different layers for different tasks also result in missing information for the main task in the low-level stage. Therefore, we propose a two-stream encoder for two tasks and align the saliency maps with the contour maps, using attention mechanisms to enhance and refine the contours of the saliency maps. Figure 3 shows the edge maps obtained from TSSTNet learning compared with CTDNet [12].

In summary, the main contributions of this paper are as follows:

(1) For the SOD of no-service rail surface detection, we propose a supervised deep learning network named TSSTNet and innovatively introduce the transformer as the backbone of the network;
(2) A two-stream encoder and a three-stream decoder are proposed to eliminate the adverse effects between tasks;
(3) A contour alignment module is presented to connect multitasking and reduce the noise at the edges of the saliency maps;
(4) A multi-feature fusion module is proposed to converge the feature maps in the three different streams of the decoder;
(5) We conducted a comparison experiment on the NRSD-MN [2] dataset with 10 SOTA methods [13,14,16–23]. The results indicate that our network performs better than the other networks on five metrics.

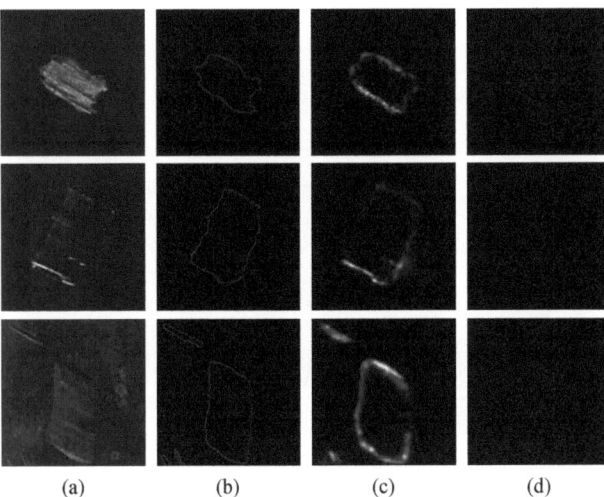

Figure 3. (**a**) No-service rail surface defects. (**b**) Edge ground truth. (**c**) Contour maps of TSSTNet. (**d**) Contour maps of CTDNet [12].

2. Related Works

2.1. Detection of Rail Defects

Rail defect detection is an integral part of the rail production process. Conventional non-destructive detection methods include magnetic particle detection, radiographic detection, eddy current detection, and ultrasonic detection. Antipov et al. [24] performed 3D computer simulations of magnetic flux leakage around transverse cracks in the rail head to detect the main characteristics of defects. Jian et al. [25] came up with an AE-signal-based detection system to detect defects by comparing the time intervals of AE wavelets. Mehel-Saidi et al. [26] proposed a method using a non-contact eddy current sensor to identify the different noises at the defect. Shi et al. [27] proposed a guided wave mode selection, which locates defects based on the different sensitivity of different modes to defects at various locations.

Meanwhile, neural-network-based vision algorithms are also popular in the field of rail defect detection. Zhang et al. [28] proposed a dual-stream neural network—one stream for generating samples and the other for classification. Zhang et al. [29] proposed an improved single-shot multibox detector (SSD) and You Only Look Once version 3 (YOLOv3), implementing the use of two networks to identify three different types of defects in parallel. Meng et al. [30] proposed a neural network framework for multitask learning to aid in track crack detection through track object detection.

2.2. Saliency Detection in RGB Images

Saliency detection, as the first step of many visual tasks, is growing in importance. It uses a computer vision algorithm and neural network learning to locate the most "remarkable" areas of the image. It helps people to highlight the areas of the image that should receive the most focus.

U-Net [31] effectively combines multilevel features using its unique U-shaped structure and skipping connection, making it the basic structure of most networks. EGNet [32] incorporates a model that obtains boundary information using low-level and high-level features, and then models the boundary information and target information. PiCANet [33] consists of an attention mechanism using pixel-wise contextual messages to learn location information for each pixel. Pyramid-Feature-Attention Network [34] consists of a context-aware pyramid feature extraction approach. U2Net [35] consists of a two-level nested U-structure and a residual U-block to capture more contextual information from different scales. ASNet [36] consists of an attention mechanism to imitate human visual attention mechanisms. CAGNet [37] consists of a feature guidance network to reduce the impacts of "salient like" appearance. PoolNet [38] consists of a global guidance module using different-sized pooling kernels to capture local and global information.

2.3. Contour Information Learning

Contour information as a separator between the foreground and background is important in saliency detection. Adding contour information to the network can significantly increase the effectiveness of saliency detection. The contour map can also refine the pixel distribution between the foreground and background. Contour detection is also increasingly being studied as a separate task.

CTDNet [12] consists of a trilateral decoder with spatial, semantic, and boundary paths. C2SNet [13] tries to graft a new branch onto a well-trained contour detection network and combines the contour task with the saliency task. PsiNet [14] consists of a structure with three parallel encoders, and one of them is used to perform the auxiliary tasks of contour detection. ENFNet [39] consists of a novel edge-guided structure to solve the problem of blurred edges caused by pooling operations.

2.4. Transformer

Bahdanau et al. [39] first applied an attention mechanism to the field of NLP. Vaswani et al. [40] first proposed a pure transformer, completely abandoning network structures such as traditional RNNs and CNNs. The transformer contains only the attention mechanism but achieves good results. After that, ViT [10] was proposed to use attention in the field of image processing. This approach splits the input image into several patches and sends them to transformer-like word vectors. Today, more and more transformers are being proposed to solve image classification, semantic segmentation and so on.

T2T [41] introduced tokens to tokens, enabling it to perform better on small datasets. DeiT [42] distills knowledge based on tokens, enabling it to perform better without pretraining on large datasets. Swin Transformer [11] consists of a multi-head self-attention mechanism based on shifted windows. PVT [43] consists of a shrinking pyramid and can contribute to downstream tasks, similar to the ResNet [44] backbone. CvT [45] combines the advantages of CNNs and transformers. It has both the dynamic attention mechanism and global modelling capabilities of transformers and the local capture capabilities of CNNs.

3. Methodology

3.1. Overview of the TSSTNet Framework

The proposed network TSSTNet is displayed in Figure 4. It comprises a two-stream encoder and a three-stream decoder. A contour alignment module (CAM) and a contour enhancement module (CEM) are proposed to use contour information learned by the edge stream to assist with saliency detection tasks. A feature fusion module (FFM) and a multi-feature fusion module (MFFM) are proposed to incorporate multi-model features. The details of the above four modules are showcased in Figure 5.

Specifically, the images are fed into two separately trained encoders after preprocessing, i.e., the edge stream and the saliency stream. Afterwards, the obtained features are fed into three different decoders—namely, the edge stream, the saliency stream, and the fusion stream—and the features from the three decoders are finally fused and output by the MFFM. The details are presented below.

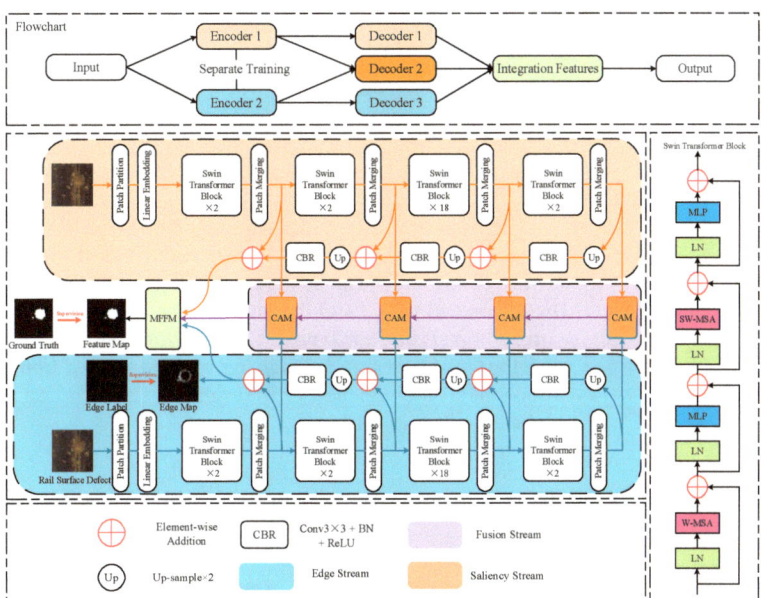

Figure 4. Structural diagrams of our proposed network TSSTNet.

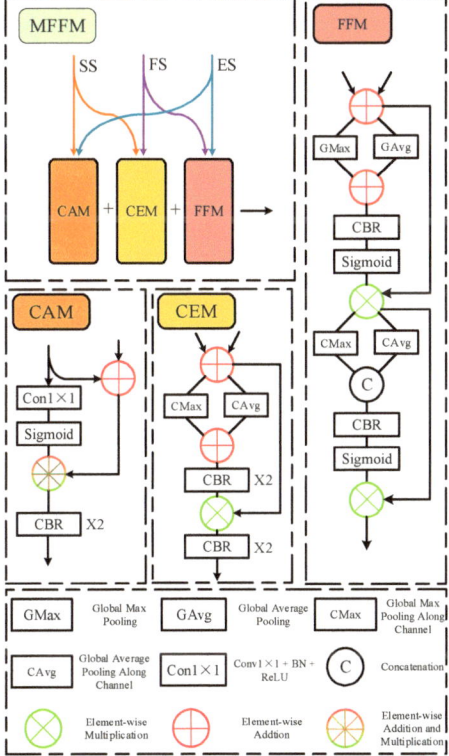

Figure 5. Structural diagrams of the contour alignment module (CAM), contour enhancement module (CEM), feature fusion module (FFM), and multi-feature fusion module (MFFM).

3.2. Two-Stream Encoder and Three-Stream Decoder

In our experiments, we found that when using a single-stream encoder to perform multitask learning, the secondary task may interfere with the primary mission. Therefore, in our network, we propose a two-stream decoder—one stream for extracting features, and the other for boundary information. Separate training parameters are used for the two streams. The experimental results prove that the two-stream encoder works better than the single-stream encoder.

We propose a three-stream decoder consisting of a saliency stream (SS), edge stream (ES), and fusion stream (FS). It makes full use of the extracted multilevel feature maps. Features from stage i can be denoted as $\{S_i\}_i^4$ in the SS, $\{E_i\}_i^4$ in the ES, and $\{F_i\}_i^4$ in the FS. The features in the SS and ES are then fused hierarchically by upsampling, convolution with a 3×3 kernel, batch normalization, and the *ReLU* activation function. At the same time, the CAM transfers the edge information learned from the ES to the SS by calculating the spatial position relationships, resulting in a precise contour. The three-stream decoder can be described as follows:

$$\begin{cases} S_i = CBR(Up(S_{i-1})) + S_i \\ E_i = CBR(Up(E_{i-1})) + E_i \end{cases} \quad (1)$$

$$F_i = CAM(S_i, E_i) + F_{i-1} \quad (2)$$

$$Result = MFFM(S_4, E_4, F_4) \quad (3)$$

where *CBR* denotes the 3×3 convolution, batch normalization, and *ReLU* function, while *Up* represents upsampling $\times 2$.

3.3. Swin Transformer Backbone

Transformers increasingly perform better than CNNs on a wide range of visual tasks. Swin Transformer [11] not only retains the advantages of the ViT [10]—such as versatility, the ability to model the global long-range dependency features, and parallel processing capability—but also incorporates CNNs' advantages of translational invariance and localization. Meanwhile, Swin Transformer [11] effectively solves the problem of heavy calculation caused by the self-attention operation. The contents of the Swin Transformer [11] are presented in Figure 4.

Specifically, Swin Transformer [11] firstly divides the RGB image into some non-overlapping patches via a patch partition operation, and then it applies a linear embedding layer on the patches, which transforms the data to a specific channel using a fully connected layer. After that, the patches are put into two successive Swin Transformer [11] blocks to extract multilevel features. The structure of the blocks is illustrated in Figure 4. With the network growing deeper, patch merging layers perform downsampling operations and reduce the resolution. Finally, we can capture the feature maps with the size of H/32, W/32, where H $\times$ W is the shape of the defect pictures. We chose Swin-B [11] as the pre-trained model in our network. Specifically, the fully connected layer converts the number of channels in the input patches to 128, while the number of repetitions of Swin Transformer [11] blocks is {2, 2, 18, 2}.

3.4. Multi-Feature Fusion Module

At the end of the network, we present a multi-feature fusion module (MFFM) that combines the features from the SS, ES, and FS two by two. For the multi-model maps in different streams, it contains three different models. These allow the network to learn boundary information while retaining the global context learned by the SS. The exact structures of the modules are described in more detail below, and the diagram of the structure of all modules is shown in Figure 5.

3.4.1. Contour Alignment Module

To use edge information to calibrate saliency detection, we propose a contour alignment module (CAM). We transform the edge map into probabilities of corresponding positions using the sigmoid function. Then, we align the edges of the rail defects by multiplying the saliency maps and the contour maps to deepen the image edges and reduce the noise between the background and foreground.

Specifically, we first merge the edge and the saliency maps by addition to form the fusion maps, and then we calculate the contour attention using a 1×1 convolution and sigmoid function. Then, we multiply and add the contour attention and the fusion map. Finally, two CBR functions are used to increase the learnability and complexity. Through contour attention, we expect the network to be more attentive to the edges of rail defects so as to achieve higher accuracy. The overall framework is shown in Figure 5. It can be formulized as follows:

$$CBR^2(x) = CBR(CBR(x)) \tag{4}$$

$$CGM(S_i, E_i) = CBR^2[Sig(Cov1(E_i)) * (E_i + S_i) + (E_i + S_i)] \tag{5}$$

where $Cov1$ denotes 1×1 convolution, Sig represents the sigmoid function, * denotes element-wise multiplication, and CBR^2 denotes two consecutive CBR functions.

3.4.2. Contour Enhancement Module and Feature Fusion Module

To combine the features in the ES and FS, we propose a contour enhancement module. Considering that the CAM aligns the contour maps to the saliency maps, we expect to let the edge maps guide the fine-tuning of the fusion maps through channel attention. Channel attention can improve the feature presentation of feature maps, while simultaneously reducing the noise at the edges of the feature maps due to the blurring of the edge maps. CEM can be presented as follows:

$$f_i = E_i + F_i \tag{6}$$

$$CEM(f_i) = CBR^2\left[CBR^2(CMax(f_i) + CAvg(f_i)) * f_i\right] \tag{7}$$

where $CMax$ represents the global max pooling operation along the channels, $CAvg$ represents the global average pooling along the channels, and f_i is an intermediate feature. In this model, the contour on the feature is enhanced again.

Inspired by [46], we propose FFM by combining channel attention and spatial attention. However, we remove the bottleneck structure of the MLP in order to reduce the amount of computation in the network. We expect the fusion maps to pass contour information to saliency maps, while the saliency maps can retain semantic context information in the fusion maps. FFM can be denoted as follows:

$$S\_A = Sig(CBR(GMax(S_i + F_i) + GAvg(S_i + F_i))) * (S_i + F_i) \tag{8}$$

$$FFM(f_i) = Sig(CBR(Cat(CMax(S\_A), CAvg(S\_A)))) * S\_A \tag{9}$$

where $GMax$ denotes the global max pooling operation along the axis, $GAvg$ denotes the global average pooling operation along the axis, S_A represents spatial attention, and Cat represents concatenation.

3.5. Training

We chose the NRSD-MN [2] dataset to train, validate, and test our network. The input size was processed to $384 \times 384 \times 3$. This also corresponds to the input size of Swin-B. We enhanced the data using random flipping, rotation, and border clipping. Swin-B was chosen to initialize the parameters of the network. The batch size was programmed to 8, and the training epochs were set at 50. An Adam optimizer was introduced to train our network. The learning rate was initialized to 5×10^{-5}, and then it decayed to 5×10^{-6} when the number of training epochs reached 30. We set the gradient clipping margin to 0.5. TSSTNet was trained on a machine with a single NVIDIA RTX 3090 and 24 GB graphics memory, and the approximate training time was 6 h. All code was implemented in the PyTorch framework.

4. Experimental Section

4.1. Datasets

To train TSSTNet, we selected the NSRD-MN [2] dataset. Zhang et al. [2] built a filming system consisting of a binocular color line-array camera, two light sources with a linear shape, and a motion transmission system. The color line-array camera reduces the light requirement to linear, uniform light. We used two linear light sources to provide linear uniform light. The camera cannot capture the entire surface of the rail at once, so a moving platform located underneath the rail carries the rail in slow motion. The specific structure is shown in Figure 6. The binocular line-array camera takes on the task of photographing. After the shoot, it manually annotates the images under the guidance of professionals from steel-testing companies. Then, we transform the annotations of the training and validation photos into edge annotations using the Canny [47] algorithm.

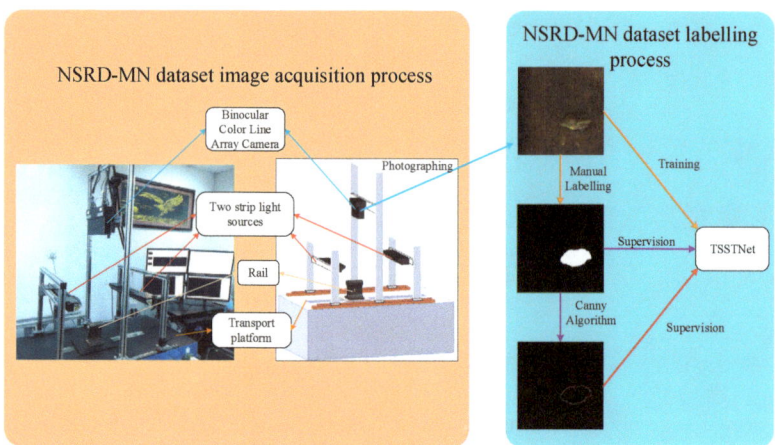

Figure 6. The process of creating the NSRD-MN dataset.

In summary, we obtained 3936 craft images of no-service rail surface defects (NRSDs), including 2158 images aged by a rust-promoting reagent, 1778 unaged pictures, and 165 natural NRSD photos. The natural set includes 115 highly similar and imitative images and 50 real images without any processing. They use metal to create scratches on the surface of the rails or use rust-promoting reagents on the surface of the metal to cause the metal to age and rust, making the manmade dataset look very similar to the natural one. The craft images were split into groups of 2086, 885, and 965 for training, validation, and testing, respectively, and the natural images were used as the test set.

4.2. Evaluation Metrics and Loss Function

To assess the capability of TSSTNet, we used five evaluation metrics commonly utilized in the field of SOD. Firstly, we chose the mean absolute error (*MAE*) [48] as our key metric, which visually shows the error between the predicted and true values. The formula is shown below:

$$MAE = \frac{1}{H \times W} \sum_{j=1}^{H} \sum_{k=1}^{W} |S(j,k) - G(j,k)| \qquad (10)$$

where *S* is the saliency map, *G* is the ground truth (GT), *j*, *k* is the location of the pixel, and H × W is the size of the entry image.

The mean F-measure (mF_β) [49] was used to demonstrate the performance of the model by calculating *Precision* and *Recall*. The weighted F-measure (wF_β) was used to evaluate the positional accuracy of the salient results. Conventionally, we set β^2 to 0.3.

$$Precision = \frac{TP}{TP + FP} \quad (11)$$

$$Recall = \frac{TP}{TP + FN} \quad (12)$$

$$F_\beta = \frac{(1+\beta^2) \times Precision \times Recall}{\beta^2 \times Precision \times Recall} \quad (13)$$

where *TP* means the true positive, *FP* means the false positive, and *FN* means the false negative in the contradiction matrix.

Structure-measure (S_α) [50] was used to measure the structural similarities between the predicted images and the annotations. There, we set α to 0.5.

$$S_\alpha = \alpha * S_o + (1-\alpha) * S_r \quad (14)$$

Enhanced-alignment measure (E_ξ) [51] focuses on the link between image-level data and local pixels by combining global pixel averages and local pixels. Finally, we also drew precision–recall (*PR*) and F-measure curves to show all information.

We choose binary cross-entropy loss (*BCELoss*) and intersection-over-union loss (*IouLoss*) as our loss functions. For training the edge maps, *BCELoss* was adopted as the edge stream loss (L_e).

$$L_e(P,G) = -\frac{1}{H \times W} \sum_{j=1}^{H} \sum_{k=1}^{W} [G(j,k) * log(P(j,k)) + (1-G(j,k)) * log(1-P(j,k))] \quad (15)$$

where *G* denotes the ground truth map, *P* denotes the prediction map, and *j, k* denotes the location of the pixel.

For training of the saliency maps, an integrated loss function (L_s) was introduced. In accordance with CTDNet [12], we set β to 0.6.

$$L_{iou} = 1 - \frac{\sum_{j=1}^{H} \sum_{k=1}^{W} G(j,k)P(j,k)}{\sum_{j=1}^{H} \sum_{k=1}^{W} [G(j,k) + P(j,k) - G(j,k) \times P(j,k)]} \quad (16)$$

$$L_s = \beta \times L_e + L_{iou} \quad (17)$$

As with the F-measure, we also chose *PR* curves to represent the relationship between precision and recall in the network results. Specifically, the precision and recall expressions indicate the percentage of true positives in the confusion matrix. The *PR* curve represents the relationship between precision and recall, and it is usually a convex curve. The higher the curve is to the right, the more effective the network. If two *PR* curves intersect, the further to the right the point is when *P* = *R*, the better the network.

4.3. Comparison of Method Performance

We compared the performance of 10 SOTA deep learning neural networks widely used in SOD tasks on the NRSD-MN dataset (i.e., BASNet [44], BSANet [45], C2FNet [46], CTDNet [12], EGNet [24], F3Net [47], PFPN [48], PiCANet [25], PoolNet+ [49], and TRACER [51]). The results are shown in Figure 7, and we have marked the experimental results of TSSTNet with a dashed box. It can be clearly seen that the salient detection results obtained by our model are more accurate. Meanwhile, we compared the networks with similar test results in more detail. The comparative diagram is shown in Figure 8. In the diagram, we compare the visualization results in terms of detection completeness and edge refinement, using different-colored rectangular boxes to box them out. Benefiting from the edge stream, contour alignment module (CAM), and multi-feature fusion module (MFFM), we obtained a clearer contour. The code is available at https://github.com/VDT-2048/TSSTNet, which is accessed on 1 October 2022.

In terms of specific evaluation results, our network shows an improved effect compared to the latest proposed network TRACER [51], in the following ways: 0.6% lower MAE, 2.6% higher mF_β, 2.8% higher wF_β, 2.0% higher S_α, and 0.8% higher E_ξ on the natural surface defects dataset (Real); and 0.3% lower MAE, 3.7% higher mF_β, 3.6% higher wF_β, 1.4% higher S_α, and 1.2% higher for E_ξ on the manmade surface defects dataset (Craft). Other comparative results are illustrated in Table 1.

The experimental results show that our network is highly competitive in saliency detection. Benefiting from the Swin Transformer [11] backbone and the two-stream multitasking encoder, our network is more accurate in feature extraction. At the same time, the CAM helps the network to distinguish between ambiguous foregrounds and backgrounds very well.

We drew PR curves and F-measure curves to show the relationship between precision and recall, as shown in Figures 9 and 10, respectively. From the PR curve analysis, we can infer that TSSTNet performs much better than the other comparison networks on the real dataset. On the craft dataset, several networks achieved similar results, but TSSTNet was still superior to the other networks.

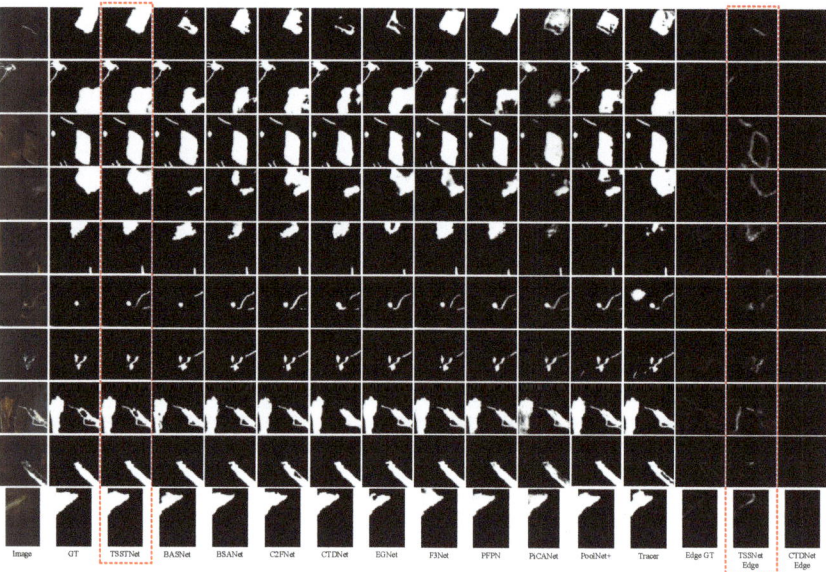

Figure 7. Visualized saliency maps of no-service rail surface defect detection compared with 10 other networks.

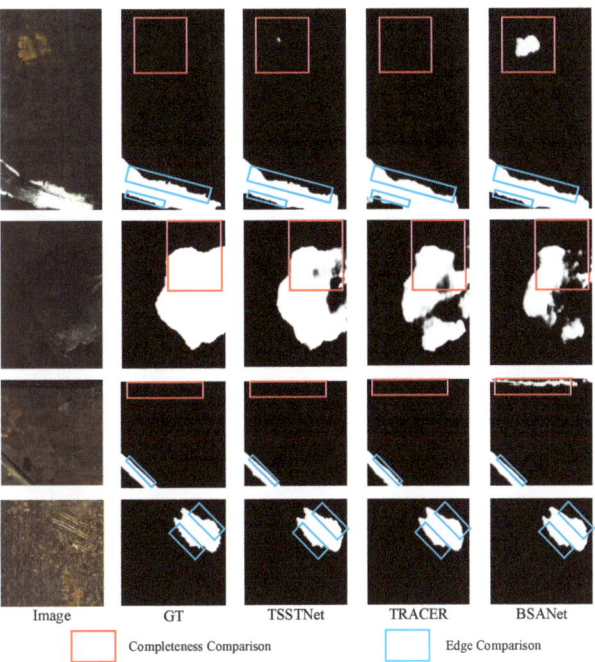

Image   GT   TSSTNet   TRACER   BSANet

☐ Completeness Comparison   ☐ Edge Comparison

Figure 8. Visual comparison of the two networks with the most similar test results from detection of completeness and edge refinement.

Table 1. Comparison with 10 other common SOD networks; ↑ means the higher the better, ↓ means the lower the better. The best two results are tagged in red and green, respectively. *MAE* is the key metric.

Methods	NRSD-MN Dataset									
	Real					Craft				
	$MAE\downarrow$	$mF_\beta\uparrow$	$wF_\beta\uparrow$	$S_\alpha\uparrow$	$E_\xi\uparrow$	$MAE\downarrow$	$mF_\beta\uparrow$	$wF_\beta\uparrow$	$S_\alpha\uparrow$	$E_\xi\uparrow$
BASNet	0.065	0.748	0.730	0.797	0.830	0.021	0.802	0.775	0.866	0.944
BSANet	0.064	0.761	0.740	0.808	0.837	0.017	0.844	0.820	0.884	0.958
C2FNet	0.063	0.761	0.705	0.805	0.850	0.021	0.817	0.738	0.859	0.949
CTDNet	0.068	0.734	0.708	0.779	0.828	0.020	0.808	0.779	0.865	0.948
EGNet	0.063	0.746	0.723	0.798	0.840	0.019	0.814	0.797	0.872	0.948
F3Net	0.060	0.771	0.754	0.822	0.847	0.018	0.824	0.799	0.879	0.950
PFPN	0.059	0.759	0.742	0.819	0.857	0.019	0.798	0.793	0.871	0.940
PiCANet	0.076	0.679	0.633	0.749	0.826	0.031	0.718	0.695	0.819	0.901
PoolNet+	0.061	0.760	0.740	0.811	0.839	0.017	0.825	0.805	0.875	0.953
TRACER	0.058	0.772	0.753	0.819	0.859	0.019	0.825	0.805	0.875	0.953
Ours	0.052	0.798	0.781	0.839	0.867	0.016	0.841	0.816	0.883	0.958

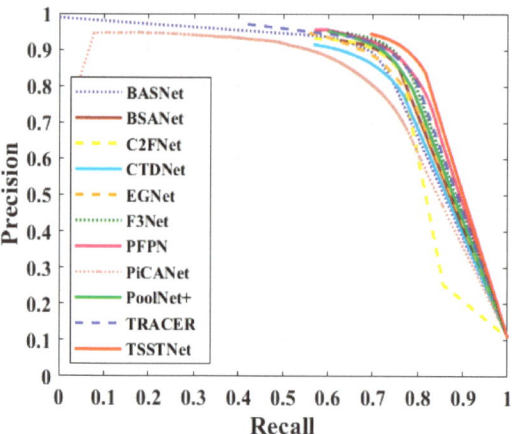

Figure 9. PR curves for the real NRSD-MN dataset.

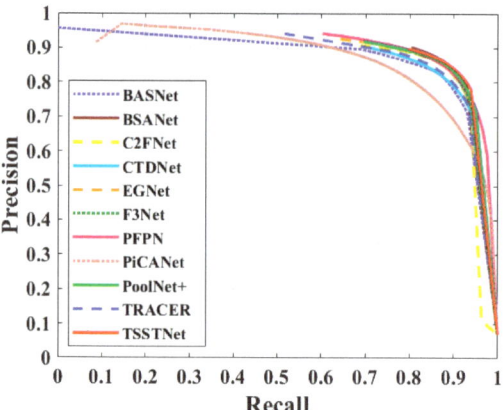

Figure 10. PR curves for the craft NRSD-MN dataset.

4.4. Ablation Experiments

To verify our conjecture about the benefits of two-stream networks, and to evaluate the advantages and disadvantages of the proposed modules, we conducted several ablation experiments. We used upsampling operations and element-wise addition to merge the hierarchical features to form a U-shaped structure. We also removed the MFFM and the FS separately using the control variable method. In the experiments to eliminate the MFFM, we used element-wise addition instead of the MFFM to fuse the ES, SS, and FS. The results of the investigation are displayed in Table 2.

The experimental results indicate that the two-stream encoder can indeed solve the problem of the auxiliary task interacting with the main task in the single-stream encoder. Furthermore, the combination of the three-stream features extracted by the two-stream encoder through the MFFM can solve the problem of conflict when different features are fused.

Table 2. Results of the ablation experiments; ↑ means the higher the better, ↓ means the lower the better. The best two results are tagged in red and green, respectively. *MAE* is the key metric.

	Swin Transformer Backbone	Two-Stream Encoder	Three-Stream Decoder	MFFM	NRSD-MN Dataset	$MAE \downarrow$	$mF_\beta \uparrow$	$wF_\beta \uparrow$	$S_\alpha \uparrow$	$E_\xi \uparrow$
Test1	✓				Real	0.056	0.787	0.767	0.825	0.856
Test2	✓	✓			Real	0.053	0.805	0.776	0.820	0.874
Test3	✓	✓	✓		Real	0.099	0.705	0.359	0.732	0.835
TSSTNet	✓	✓	✓	✓	Real	0.052	0.798	0.781	0.839	0.867
Test4	✓				Craft	0.017	0.831	0.807	0.880	0.955
Test5	✓	✓			Craft	0.020	0.833	0.810	0.860	0.954
Test6	✓	✓	✓		Craft	0.062	0.761	0.389	0.771	0.921
TSSTNet	✓	✓	✓	✓	Craft	0.016	0.841	0.816	0.883	0.958

5. Conclusions

In this paper, we present a multi-stream neural network named Two-Stream Swin Transformer Network (TSSTNet) to meet the challenge of blurring the edges of rail surface defects and distinguishing the foreground from the background. TSSTNet uses two separately trained encoders to extract saliency features and edge features, which is a good solution to the problem of inter-task interference when a single-stream encoder is working on multiple tasks. This also makes TSSTNet highly capable of edge refinement. At the same time, we propose a contour alignment module (CAM) to use spatial attention to fuse features from different streams, which calibrates the edges of the saliency detection map, reduces noise at the foreground–background junction, and helps the network to locate defects. Moreover, a multi-feature fusion module (MFFM) is proposed to solve the problem of conflicting features at different levels of the three-stream decoder, which is able to reduce the variability in the fusion of features learned from different streams. However, a dual-stream decoder with separately trained parameters would result in a larger model and more computational effort. In subsequent studies, we will remove the less important parts of the network by pruning, compression, and other operations to lighten the network. In addition to this, we will commit to the use of multimodal information—including RGB-D and RGB-T images [51]—for the detection of no-service rail surface defects.

Author Contributions: Conceptualization, C.W. and S.M.; methodology, C.W.; software, C.W. and S.M.; validation, C.W.; formal analysis, C.W.; investigation, C.W.; resources, S.M. and K.S.; data curation, C.W.; writing—original draft preparation, C.W., S.M. and K.S.; writing—review and editing, C.W., S.M. and K.S.; visualization, C.W.; supervision, K.S.; project administration, K.S.; funding acquisition, K.S. All authors have read and agreed to the published version of the manuscript.

Funding: This work is supported by the National Natural Science Foundation of China (51805078), the Fundamental Research Funds for the Central Universities (N2103011), the Central Guidance on Local Science and Technology Development Fund (2022JH6/100100023).

Institutional Review Board Statement: Not applicable.

Informed Consent Statement: Not applicable.

Data Availability Statement: Not applicable.

Conflicts of Interest: The authors declare no conflict of interest.

References

1. Gan, J.; Li, Q.; Wang, J.; Yu, H. A hierarchical extractor-based visual rail surface inspection system. *IEEE Sens. J.* **2017**, *17*, 7935–7944. [CrossRef]
2. Zhang, D.; Song, K.; Xu, J.; He, Y.; Niu, M.; Yan, Y. MCnet: Multiple Context Information Segmentation Network of No-Service Rail Surface Defects. *IEEE Trans. Instrum. Meas.* **2021**, *70*, 1–9. [CrossRef]
3. Xu, P.; Zeng, H.; Qian, T.; Liu, L. Research on defect detection of high-speed rail based on multi-frequency excitation composite electromagnetic method. *Measurement* **2022**, *187*, 110351. [CrossRef]

4. Hao, F.; Shi, J.; Zhang, Z.; Chen, R.; Zhu, S. Canny edge detection enhancement by general auto-regression model and bi-dimensional maximum conditional entropy. *Optik* **2014**, *125*, 3946–3953. [CrossRef]
5. Cao, B.; Li, J.; Liu, C.; Qin, L. Defect detection of nickel plated punched steel strip based on improved least square method. *Optik* **2020**, *206*, 164331. [CrossRef]
6. Hao, Q.; Zhang, X.; Wang, Y.; Shen, Y.; Makis, V. A novel rail defect detection method based on undecimated lifting wavelet packet transform and Shannon entropy-improved adaptive line enhancer. *J. Sound Vib.* **2018**, *425*, 208–220. [CrossRef]
7. Shakeel, M.S.; Zhang, Y.; Wang, X.; Kang, W.; Mahmood, A. Multi-scale Attention Guided Network for End-to-End Face Alignment and Recognition. *J. Vis. Commun. Image Represent.* **2022**, *88*, 103628. [CrossRef]
8. Baffour, A.A.; Qin, Z.; Wang, Y.; Qin, Z.; Choo, K.K.R. Spatial self-attention network with self-attention distillation for fine-grained image recognition. *J. Vis. Commun. Image Represent.* **2021**, *81*, 103368. [CrossRef]
9. Borji, A.; Cheng, M.; Hou, Q.; Jiang, H.; Li, J. Salient object detection: A survey. *Comput. Vis. Media* **2019**, *5*, 117–150. [CrossRef]
10. Dosovitskiy, A.; Beyer, L.; Kolesnikov, A.; Weissenborn, D.; Zhai, X.; Unterthiner, T.; Dehghani, M.; Minderer, M.; Heigold, G.; Gelly, S.; et al. An image is worth 16x16 words: Transformers for image recognition at scale. *arXiv* **2020**, arXiv:2010.11929.
11. Liu, Z.; Lin, Y.; Cao, Y.; Hu, H.; Wei, Y.; Zhang, Z.; Lin, S.; Guo, B. Swin transformer: Hierarchical vision transformer using shifted windows. In Proceedings of the IEEE/CVF International Conference on Computer Vision, Montreal, QC, Canada, 10–17 October 2021; pp. 10012–10022.
12. Zhao, Z.; Xia, C.; Xie, C.; Li, J. Complementary trilateral decoder for fast and accurate salient object detection. In Proceedings of the 29th ACM International Conference on Multimedia, Virtual, 20–24 October 2021; pp. 4967–4975.
13. Li, X.; Yang, F.; Cheng, H.; Liu, W. Contour knowledge transfer for salient object detection. In Proceedings of the European Conference on Computer Vision (ECCV), Munich, Germany, 8–14 September 2018; pp. 355–370.
14. Murugesan, B.; Sarveswaran, K.; Shankaranarayana, S.M.; Ram, K.; Joseph, J.; Sivaprakasam, M. Psi-Net: Shape and boundary aware joint multi-task deep network for medical image segmentation. In Proceedings of the 2019 41st Annual International Conference of the IEEE Engineering in Medicine and Biology Society (EMBC), Berlin, Germany, 23–27 July 2019; pp. 7223–7226. [CrossRef]
15. Tu, Z.; Ma, Y.; Li, C.; Tang, J.; Luo, B. Edge-guided non-local fully convolutional network for salient object detection. *IEEE Trans. Circuits Syst. Video Technol.* **2020**, *31*, 582–593. [CrossRef]
16. Zhang, J.; Li, S.; Yan, Y.; Ni, Z.; Ni, H. Surface Defect Classification of Steel Strip with Few Samples Based on Dual-Stream Neural Network. *Steel Res. Int.* **2022**, *93*, 2100554. [CrossRef]
17. Wu, H.; Xiao, B.; Codella, N.; Liu, M.; Dai, X.; Yuan, L.; Zhang, L. Cvt: Introducing convolutions to vision transformers. In Proceedings of the IEEE/CVF International Conference on Computer Vision, Montreal, QC, Canada, 10–17 October 2021; pp. 22–31.
18. Woo, S.; Park, J.; Lee, J.Y.; Kweon, I.S. Cbam: Convolutional block attention, module. In Proceedings of the European Conference on Computer Vision (ECCV), Munich, Germany, 8–14 September 2018; pp. 3–19.
19. Ding, L.; Goshtasby, A. On the Canny edge detector. *Pattern Recognit.* **2001**, *34*, 721–725. [CrossRef]
20. Willmott, C.J.; Matsuura, K. Advantages of the mean absolute error (MAE) over the root mean square error (RMSE) in assessing average model performance. *Clim. Res.* **2005**, *30*, 79–82. [CrossRef]
21. Achanta, R.; Hemami, S.; Estrada, F.; Susstrunk, S. Frequency-tuned salient region detection. In Proceedings of the 2009 IEEE Conference on Computer Vision and Pattern Recognition, Miami, FL, USA, 20–25 June 2009; pp. 1597–1604. [CrossRef]
22. Fan, D.P.; Cheng, M.M.; Liu, Y.; Li, T.; Borji, A. Structure-measure: A new way to evaluate foreground maps. In Proceedings of the IEEE International Conference on Computer Vision, Venice, Italy, 22–29 October 2017; pp. 4548–4557.
23. Fan, D.P.; Gong, C.; Cao, Y.; Ren, B.; Cheng, M.; Borji, A. Enhanced-alignment measure for binary foreground map evaluation. *arXiv* **2018**, arXiv:1805.10421.
24. Antipov, A.G.; Markov, A.A. Detectability of Rail Defects by Magnetic Flux Leakage Method. *Russ. J. Nondestruct. Test.* **2019**, *55*, 277–285. [CrossRef]
25. Jian, H.; Lee, H.R.; Ahn, J.H. Detection of bearing/rail defects for linear motion stage using acoustic emission. *Int. J. Precis. Eng. Manuf.* **2013**, *14*, 2043–2046. [CrossRef]
26. Mehel-Saidi, Z.; Bloch, G.; Aknin, P. A subspace method for detection and classification of rail defects. In Proceedings of the 2008 16th European Signal Processing Conference, Lausanne, Switzerland, 25–29 August 2008; pp. 1–5.
27. Shi, H.; Zhuang, L.; Xu, X.; Yu, Z.; Zhu, L. An Ultrasonic Guided Wave Mode Selection and Excitation Method in Rail Defect Detection. *Appl. Sci.* **2019**, *9*, 1170. [CrossRef]
28. Zhang, H.; Song, Y.; Chen, Y.; Zhong, H. MRSDI-CNN: Multi-model rail surface defect inspection system based on convolutional neural networks. *IEEE Trans. Intell. Transp. Syst.* **2021**, *23*, 11162–11177. [CrossRef]
29. Meng, S.; Kuang, S.; Ma, Z.; Wu, Y. MtlrNet: An Effective Deep Multitask Learning Architecture for Rail Crack Detection. *IEEE Trans. Instrum. Meas.* **2022**, *71*, 1–10. [CrossRef]
30. Ronneberger, O.; Fischer, P.; Brox, T. U-net: Convolutional networks for biomedical image segmentation. In Proceedings of the International Conference on Medical Image Computing and Computer-Assisted Intervention, Munich, Germany, 5–9 October 2015; Springer: Cham, Switzerland, 2015; pp. 234–241.
31. Zhao, J.X.; Liu, J.J.; Fan, D.P.; Cao, Y. EGNet: Edge guidance network for salient object detection. In Proceedings of the IEEE/CVF International Conference on Computer Vision, Seoul, Korea, 27 October–November 2019; pp. 8779–8788.

32. Liu, N.; Han, J.; Yang, M.H. Picanet: Learning pixel-wise contextual attention for saliency detection. In Proceedings of the IEEE conference on computer vision and pattern recognition, Salt Lake City, UT, USA, 18–23 June 2018; pp. 3089–3098.
33. Zhao, T.; Wu, X. Pyramid feature attention network for saliency detection. In Proceedings of the IEEE/CVF Conference on Computer Vision and Pattern Recognition, Long Beach, CA, USA, 16–27 June 2019; pp. 3085–3094.
34. Qin, X.; Zhang, Z.; Huang, C.; Dehghan, M.; Zaiane, O.R. U2-Net: Going deeper with nested U-structure for salient object detection. *Pattern Recognit.* **2020**, *106*, 107404. [CrossRef]
35. Zhu, L.; Feng, S.; Zhu, W.; Chen, X. ASNet: An adaptive scale network for skin lesion segmentation in dermoscopy images. In *Medical Imaging 2020: Biomedical Applications in Molecular, Structural, and Functional Imaging*; SPIE: Bellingham, WA, USA, 2020; Volume 11317, pp. 226–231.
36. Mohammadi, S.; Noori, M.; Bahri, A.; Majelan, S.G. CAGNet: Content-aware guidance for salient object detection. *Pattern Recognit.* **2020**, *103*, 107303. [CrossRef]
37. Liu, J.J.; Hou, Q.; Cheng, M.M. A simple pooling-based design for real-time salient object detection. *arXiv* **2019**, arXiv:1904.09569.
38. Bahdanau, D.; Cho, K.; Bengio, Y. Neural machine translation by jointly learning to align and translate. *arXiv* **2014**, arXiv:1409.0473.
39. Vaswani, A.; Shazeer, N.; Parmar, N.; Uszkoreit, J.; Jones, L.; Gomez, A.N.; Kaiser, L.; Polosukhin, I. Attention is all you need. *Adv. Neural Inf. Process. Syst.* **2017**, *30*. Available online: https://proceedings.neurips.cc/paper/2017/file/3f5ee243547dee91fbd053c1c4a845aa-Paper.pdf (accessed on 7 October 2022).
40. Yuan, L.; Chen, Y.; Wang, T.; Yu, W.; Shi, Y.; Jiang, Z.; Tay, F.E.H.; Feng, J.; Yan, S. Tokens-to-token vit: Training vision transformers from scratch on imagenet. In Proceedings of the IEEE/CVF International Conference on Computer Vision, Montreal, QC, Canada, 11–17 October 2021; pp. 558–567.
41. Touvron, H.; Cord, M.; Douze, M.; Massa, F.; Sablayrolles, A.; Jegou, H. Training data-efficient image transformers & distillation through attention. In Proceedings of the 38th International Conference on Machine Learning, PMLR 2021, Virtual, 18–24 July 2021; Volume 139, pp. 10347–10357.
42. Wang, W.; Xie, E.; Li, X.; Fan, D.; Song, K.; Liang, D.; Lu, T.; Luo, P. Pyramid vision transformer: A versatile backbone for dense prediction without convolutions. In Proceedings of the IEEE/CVF International Conference on Computer Vision, Montreal, QC, Canada, 11–17 October 2021; pp. 568–578.
43. He, K.; Zhang, X.; Ren, S.; Sun, J. Deep residual learning for image recognition. In Proceedings of the IEEE Conference on Computer Vision and Pattern Recognition, San Francisco, CA, USA, 18–20 June 2016; pp. 770–778.
44. Qin, X.; Zhang, Z.; Huang, C.; Gao, C.; Dehghan, M.; Jagersand, M. Basnet: Boundary-aware salient object detection. In Proceedings of the IEEE/CVF Conference on Computer Vision and Pattern Recognition, Long Beach, CA, USA, 16–17 June 2019; pp. 7479–7489.
45. Zhu, H.; Li, P.; Xie, H.; Yan, X.; Liang, D.; Chen, D.; Wei, M.; Qin, J. I can find you! Boundary-guided Separated Attention Network for Camouflaged Object Detection. AAAI 2022. Available online: https://ojs.aaai.org/index.php/AAAI/article/view/20273 (accessed on 7 October 2022).
46. Sun, Y.; Chen, G.; Zhou, T.; Zhang, Y.; Liu, N. Context-aware cross-level fusion network for camouflaged object detection. *arXiv* **2021**, arXiv:2105.12555.
47. Wei, J.; Wang, S.; Huang, Q. F³Net: Fusion, feedback and focus for salient object detection. In Proceedings of the AAAI Conference on Artificial Intelligence 2020, New York, NY, USA, 7–12 February 2020; Volume 34, pp. 12321–12328.
48. Wang, B.; Chen, Q.; Zhou, M.; Zhang, Z.; Jin, X.; Gai, K. Progressive feature polishing network for salient object detection. In Proceedings of the AAAI Conference on Artificial Intelligence 2020, New York, NY, USA, 7–12 February 2020; Volume 34, pp. 12128–12135.
49. Liu, J.; Hou, Q.; Liu, Z.; Cheng, M. PoolNet+: Exploring the Potential of Pooling for Salient Object Detection. In *IEEE Transactions on Pattern Analysis and Machine Intelligence*; IEEE: Piscataway, NJ, USA, 2022. [CrossRef]
50. Lee, M.S.; Shin, W.S.; Han, S.W. TRACER: Extreme Attention Guided Salient Object Tracing Network (Student Abstract). In Proceedings of the AAAI Conference on Artificial Intelligence, Pomona, CA, USA, 24–28 October 2022; Volume 36, pp. 12993–12994.
51. Song, K.; Wang, J.; Bao, Y.; Huang, L.; Yan, Y. A Novel Visible-Depth-Thermal Image Dataset of Salient Object Detection for Robotic Visual Perception. In *IEEE/ASME Transactions on Mechatronics*; IEEE: Piscataway, NJ, USA, 2022. [CrossRef]

Article

A Distribution-Preserving Under-Sampling Method for Imbalance Defect Recognition in Castings

Han Yu [1], Xinyue Li [1], Xingjie Li [1,2,*], Chunyu Hou [1], Shangyu Liu [1] and Huasheng Xie [1,*]

[1] State Key Laboratory of Light Alloy Foundry Technology for High-End Equipment, Shenyang Research Institute of Foundry Co., Ltd., Shenyang 110022, China

[2] National Key Laboratory for Precision Hot Processing of Metal, Harbin Institute of Technology, Harbin 150006, China

* Correspondence: lixj@chinasrif.com (X.L.); xiehs@chinasrif.com (H.X.)

Abstract: Data imbalance is a crucial factor that limits the performance of automatic defect recognition systems in castings. The bias and deterioration of the model are generated by massive normal samples and minor defect samples. Traditional re-sampling methods randomly change the data distribution and ignore the significant intra-class difference among all normal samples. Therefore, this paper proposes a distribution-preserving under-sampling method for imbalance defect-recognition in castings. In detail, our method divides all normal samples into several sub-groups by cluster analysis and reassembles them into some balance datasets, which makes the normal samples in all balance datasets have an identical distribution with the original imbalance dataset. Finally, experiments on our dataset with 3260 images indicate that the proposed method achieves a 0.816 AUC (area under curve) score, which demonstrates significant advantages compared to cost-sensitive learning and re-sampling methods.

Keywords: castings defect recognition; imbalance classification; distribution-preserving

1. Introduction

Casting is an important forming means to achieve high-efficiency and low-cost manufacture and is an irreplaceable process for metal parts with complex structure. A large number of core components are formed by the casting process in the high-end manufacturing industry. Unfortunately, castings inevitably have different internal defects because of casting materials and processes. These internal defects will seriously affect the mechanical properties of castings, even directly leading to scrap.

To obtain the internal information of castings, digital radiography (DR) technology is utilized and gradually became the first choice for nondestructive testing. DR utilizes X-ray to penetrate the castings and directly produces a digital image through the digital detector array (DDA). Then, the inspectors judge whether there is a defect in the casting by observing the change of gray level in the digital image. However, manual visual inspection not only is laborious and inefficient but also easy to be affected by the ability and experience of the inspectors. Consequently, automatic defect recognition (ADR) has collected much attention and became one of the research hot points. Early research about ADR mainly focused on handcrafted feature extraction and traditional machine learning. Mery et al. [1] utilized a single filter to track potential defects in image sequences. Hernandez et al. [2] used the neuro-fuzzy method to classify the defect and normal samples. Zhao et al. [3] proposed a defect recognition framework for automobile wheels, which employed the gray arranging pairs method to segment the defects, then their randomly distributed triangle features were extracted and fed into a sparse representation classifier to achieve defect classification. Overall, these approaches heavily rely on the handcrafted features designed by experts. Unfortunately, the handcrafted features usually have poor robustness, which fails to adapt to the change of position, structure, and ray intensity.

As deep learning develops, current ADR systems in casting mainly realize defect recognition through a convolutional neural network (CNN). Du et al. [4] used the Faster-RCNN [5] with feature pyramid networks (FPN) [6] to locate the defect. Yu et al. [7] proposed an adaptive CNN model to realize multi-class defect segmentation. A variable attention nested UNet++ network [8] was introduced for defect segmentation in the X-ray image.

The above research aims to achieve end-to-end defect location and segmentation, which can provide more comprehensive information about the defect. However, they also have a slow detection speed. In actual detection, the number of images without defects is far more than the number of images with defects. Thus, the image-level recognition method can be deployed at the front of the entire detection process to improve detection efficiency. Mery et al. [9] extensively tested the combined performance of multiple features and classifiers in the GDX-ray [10] dataset. This work crops the original image into enormous, small patches (32×32). Because of low resolution, deep learning methods perform worse than traditional machine learning. Subsequently, Mery [11] used GAN and physical simulation methods to produce more virtual defect images and boost accuracy. Tang et al. [12] combined spatial attention mechanisms and bi-linear pooling into a CNN model to increase the representation power. Similar to [12], Hu et al. [13] presented a two-stage training procedure, which firstly forces the network to classify the casting type and then to classify the defect and normal samples. Jiang et al. [14] presented the mutual-channel loss and an attention-guided data augmentation method to boost the original VGG network.

Regrettably, the above research is dedicated to designing higher performance CNN on balance datasets and sidestepping the fact of class imbalance that significantly impacts the recognition accuracy. At present, there are two main solutions to solve the imbalance classification: the re-weighting and the re-sampling method. The re-weighting method is also called cost-sensitive learning. It adopts re-weighting strategies to adjust the loss of minor class samples. Weighted cross-entropy loss is the simplest way whose weight is the inverse class frequencies. Seesaw loss [15] adaptively re-balances gradients of minor-class and major-class samples with mitigation factor and the compensation factor. Focal loss [16] is proposed to solve the class imbalance in object detection. It inversely re-weights classes by prediction probabilities, so that it can give higher weights to the minor classes but lower weights to the major classes. The re-sampling method [17,18] directly changes the data distribution and forms some balance datasets by randomly over-sampling the minority class or under-sampling the majority class.

However, the random re-sampling method ignores the intra-class difference in defect recognition. As shown in Figure 1, these normal samples are different in gray level, texture, and so on. In Figure 1a, this kind of image is brighter than others. In Figure 1b, such images have some false defects because of the concave-convex surface of castings. Figure 1c stands for the rest of the samples whose background lacks detail. Based on the previous analysis, the random under-sampling method will change the original distribution information in all normal samples and lead to the variability of balance datasets generated by re-sampling. It will break the important assumption in deep learning: independent identical distribution. Considering this, a distribution-preserving under-sampling method is proposed for imbalance defect recognition in castings. By cluster analysis, we divide all normal samples into several groups and reassemble them into some balance datasets. The normal samples in these balance datasets follow a similar distribution to the original data. Benefitting from the above improvement, our methods achieve a better accuracy under imbalanced data.

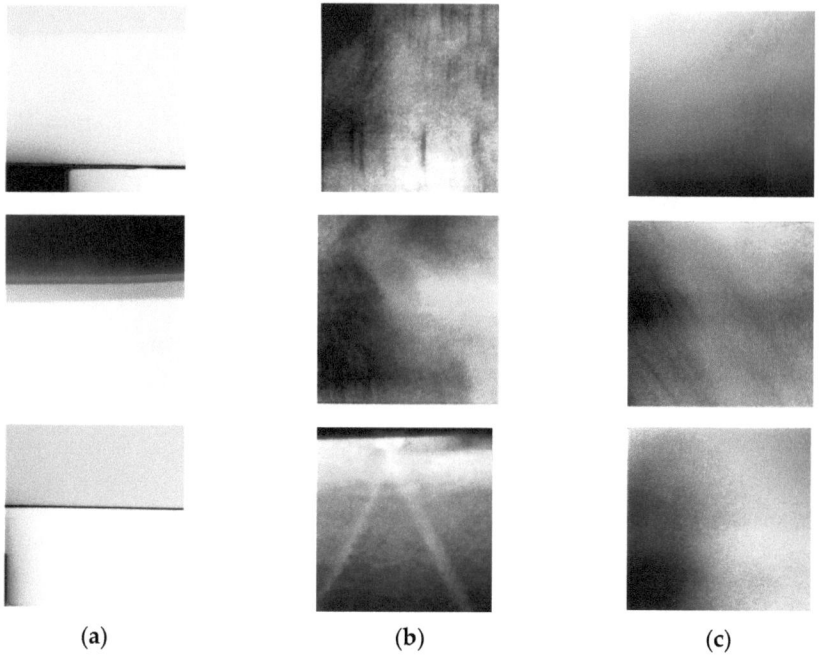

Figure 1. Three typical normal samples. (**a**): brighter normal samples (**b**): normal samples with false defects (**c**): normal samples with missing details.

2. Methods

2.1. Overview

In this section, we present our method for imbalance defect recognition. Our method firstly extracts the image features of normal samples by a pre-trained network, and then clusters them into some groups based on k-means++ [19] approach. Some normal samples belonging to each group are selected and recombined to form several new subsets. Each new subset is combined with all defect samples, resulting in some balance training datasets. We train our deep learning model by each training dataset and obtain some network weights. Through model fusion, our method produces better performance under an imbalanced dataset. The details are shown in Figure 2.

2.2. Distribution-Preserving Under-Sample

Assume that a class-imbalanced dataset $D = \{D_n \cup D_d\}$ includes normal samples (D_n) and defect samples (D_d). The numbers of the D_n, D_d are N_n, N_d. N_n is far larger than N_d. Our goal is to build some balanced datasets through under-sample methods for relieving the imbalance.

Traditional under-sample methods select some samples randomly. However, this method will change the prior distribution information in the normal class because of large intra-class differences. We believe that the subsets obtained after under-sampling should maintain the same distribution as the original dataset. Motivated by this, a distribution-preserving under-sample method is proposed to acquire multiple subsets that are identically distributed with the original normal class.

For our method, a ResNet18 [20] pre-trained by ImageNet [21] is first employed to extract the high-dimension features of normal samples because it has been proven to be a good feature extractor. These features are then PCA-reduced to 128 dimensions, whitened, and L2-normalized. K-means++, a standard clustering approach, is widely used and can be regarded as a baseline clustering method. It takes the reduced dimension features as input and clusters them into K distinct groups $(G_1, G_2, \ldots, G_k)$ based on an Euclidean distance. Each group represents the sub-class of the normal sample.

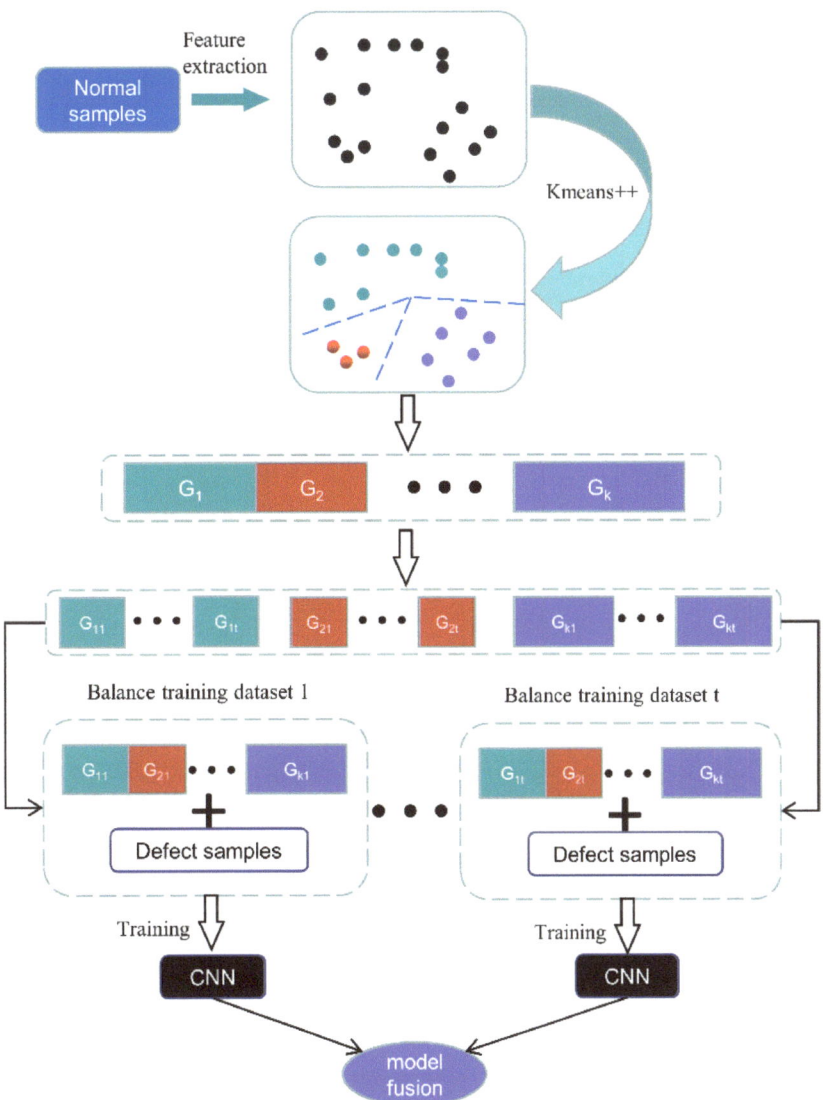

Figure 2. The pipeline of the proposed method.

To ensure the same distribution between each subset and the original dataset, we randomly select $p\%$ of the samples from each group and repeat t times, so several subgroups $(G_{11}, G_{12}, \ldots, G_{1t}, \ldots, G_{kt})$ are formed. Finally, t balanced datasets $(BD_1, BD_2, \ldots, BD_t)$ are aggregated by:

$$BD_i = \{G_{1i} \cup G_{2i} \cup, \ldots, G_{ki} \cup D_d\}, i = 1, 2, \ldots, t, \quad (1)$$

The number of normal samples in each BD can be computed as follows:

$$N_{BD} = N_n \times p\%, \quad (2)$$

In general, CNN does not need a totally balanced dataset, so we set the N_{BD} slightly larger than N_d by adjusting p.

2.3. Network

CNN has demonstrated impressive ability in many computer vision tasks. It can automatically extract general and robust features from the input data by back-propagation. CNN is usually composed of convolution, activation functions, and fully connected layers. Through careful designing of network architecture, a multiple classical CNN model is proposed. In this work, we consider the network performance and inference time comprehensively; ResNet18 and MobileNetV2 [22] are selected for defect recognition.

ResNet is used to solve the gradient disappearance with the increase in network depth. The main idea of ResNet is to introduce a residual block that forces the network to learn identity mapping rather than fitting ground truth directly. Through stacking some residual blocks, multiple versions of ResNet are produced, such as ResNet18, ResNet50 ResNet101, and so on. MobileNetV2 is improved based on ResNet to make it lighter, which proposed the inverted residual block. Different from the residual block, this module first expands the channel of input features to a high dimension and filters it with a lightweight depth-wise convolutional layer. The features are subsequently projected back to a low dimension with a 1×1 convolution. The details of the residual block and inverted residual block are shown in Figure 3.

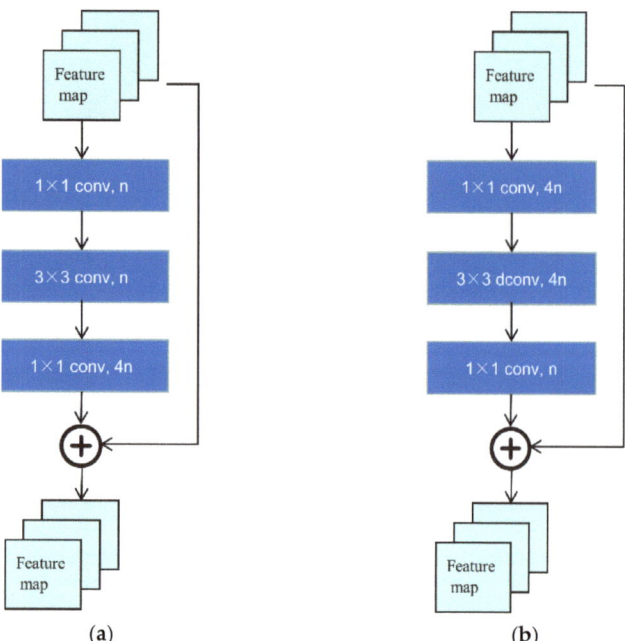

Figure 3. (**a**): residual block (**b**): inverted residual block, n means channel numbers.

In this study, ResNet18 and MobileNetV2 are modified to adapt to this task. We employ the pre-trained weight from ImageNet as an initialization and remove their all fully connected layers. Then, a new fully connected layer is added to the network. The softmax layer transfers the output of the fully connected layer into a probability distribution. Finally, we train t CNN models based on t balanced datasets.

2.4. Model Fusion

Although our under-sample method can produce several balanced and identically distributed datasets, it just includes part of the whole normal samples. The model trained by different datasets must have some bias which is harmful to generalization performance. Thus, we employ the model fusion strategy to solve it. Suppose f_i is the CNN model trained by BD_i, θ_i is the parameter of it. $f_i(x, \theta_i)$ stands for the CNN output of the sample x. By averaging the output of multiple CNN, the result y without bias can be obtained.

$$y = \sum_{i=1}^{t} f_i(x, \theta_i) \tag{3}$$

3. Experimental Setup

3.1. Datasets

We collect 3260 X-rays whose resolution is 480 × 480 of aluminum alloy castings from the line for evaluating our method. There are 630 defective samples and 2630 normal samples in the total dataset. The training dataset is imbalanced to keep consistent with the actual task for defect recognition. The evaluation and test datasets are balanced for a better comparison. Table 1 shows the detailed number. Figure 4 shows some typical normal and defect samples.

Table 1. The number of samples in this dataset.

Dataset	Normal	Defect	Total
Train	2430	430	2860
Validation	100	100	200
Test	100	100	200

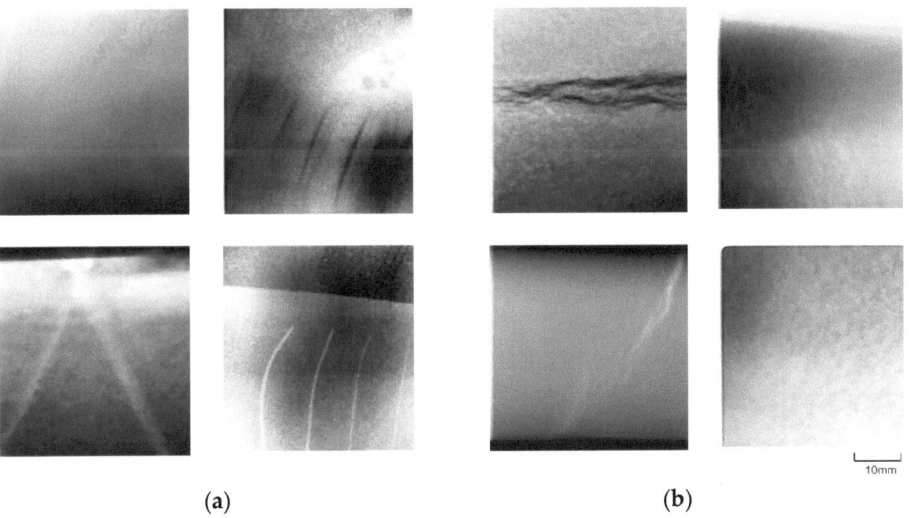

Figure 4. (**a**): normal samples (**b**): defect samples.

3.2. Implementation Details

The computation platform used in this work was Intel Core i9-9920X, 64G memory, and TITAN-RTX with 24G GPU memory. All the models were implemented by the MindSpore frame framework. The optimizer is Adam. The batch size and the learning rate were 16 and 0.001, respectively.

3.3. Evaluation Metric

Accuracy is usually used for evaluating the balance classification task. However, the tolerance for normal samples to be detected as defective samples is far greater than that for defective samples to be detected as normal samples. So, accuracy is not entirely suitable for evaluating it. In this task, AUC (area under curve) score is the main evaluation metric. AUC score is obtained by calculating the area under the ROC (receiver operating characteristic) with multi-thresholds. When the threshold is 0.5, precision and recall are also reported. The specific calculation formula of precision and recall are as follows:

$$Presicion = \frac{TP}{TP + FP} \quad (4)$$

$$Recall = \frac{TP}{TP + FN} \quad (5)$$

TP and *FN* are the numbers of defect samples classified correctly or not. *FP* represents the number of normal samples classified incorrectly.

4. Results and Discussion

4.1. Compare with Other Methods

In order to illustrate the advancement of our method, we compare it to other methods of imbalance classification including cost-sensitive learning and re-sampling method. At the same time, we test our method under two different backbone networks to prove its generalization. Every method is tested five times and the results are shown in Table 2. The baseline is the CNN model trained with standard cross-entropy loss. The baseline considers all samples as normal samples and produces the worst result due to extreme imbalance. Cost-sensitive learning improves it by giving less weight to normal samples. Simple weight cross-entropy loss performs better than focal and seesaw loss. Re-sampling methods also achieve higher accuracy. When ResNet18 is used as the backbone network, the over-sample method performs better than the under-sample, but when the backbone network is MobileNetV2, under-sample performs better than the over-sample method. In general, re-sampling methods are more suitable for our dataset than cost-sensitive learning. Our method achieves the best result under different backbone networks, especially MobileNetV2. The AUC score of our method is 3.8% higher than the under-sample method, which proves the effectiveness of our approach.

Table 2. The quantification results of different methods towards class-imbalance classification.

Backbone	Method	Precision	Recall	AUC
ResNet18	CE loss	0.5005 ± 0.001	0.996 ± 0.008	0.5509 ± 0.0735
	WCE loss	0.668 ± 0.0333	0.846 ± 0.0301	0.7437 ± 0.0291
	Focal loss	0.538 ± 0.0571	0.494 ± 0.2463	0.5613 ± 0.0435
	Seesaw loss	0.668 ± 0.0098	0.89 ± 0.0482	0.74 ± 0.0232
	Over-sample	0.6534 ± 0.0163	0.824 ± 0.0508	0.7797 ± 0.0202
	Under-sample	0.6537 ± 0.0757	0.75 ± 0.1838	0.7408 ± 0.035
	Ours	0.6326 ± 0.0227	0.84 ± 0.0562	0.7891 ± 0.0168
MobileNetV2	CE loss	0.5 ± 0.0	1 ± 0	0.5101 ± 0.075
	WCE loss	0.709 ± 0.0256	0.846 ± 0.0287	0.7864 ± 0.0355
	Focal loss	0.529 ± 0.0208	0.598 ± 0.0838	0.5408 ± 0.0194
	Seesaw loss	0.675 ± 0.037	0.868 ± 0.0343	0.7656 ± 0.0236
	Over-sample	0.6712 ± 0.0316	0.84 ± 0.0701	0.7786 ± 0.0153
	Under-sample	0.6674 ± 0.0203	0.8224 ± 0.0528	0.7859 ± 0.0069
	Ours	0.6398 ± 0.0155	0.9080 ± 0.0192	0.8158 ± 0.0160

4.2. The Influence on Different Imbalance Ratios

To prove the robustness of our method, a further experiment is conducted under different imbalance ratios (normal sample/defect sample). We gradually reduce part of the defect samples in the training dataset for simulating the more imbalanced situation. MobileNetV2 is utilized as the backbone network and part of the cost-sensitive learning methods and re-sampling methods are employed to make a comparison with our approach based on the results in 4.1. We also adjust the weight factors according to the imbalance ratios for WCE and seesaw loss methods. Figure 5 shows the AUC scores of different methods when the number of defect samples reduces to 1/2, 1/4, 1/8, and 1/16 of the original. It is consistent with the intuition that the experimental performance of most comparison methods significantly decreases as the number of defect samples decreases. Overall, the tolerance of cost-sensitive learning methods is worst for imbalanced datasets. When the defect samples are only 1/8 of the original, the AUC of seesaw loss reduces by 27.45%. This is because these methods cannot essentially deal with the issue of lacking information, particularly on limited data amounts. Re-sampling methods perform better than cost-sensitive learning methods, especially the under-sampling method. However, it still has a big gap with our method. As the number of defect samples decreases, the AUC score of our method is still optimal. It demonstrates that our method has significant advantages in tolerance for imbalance.

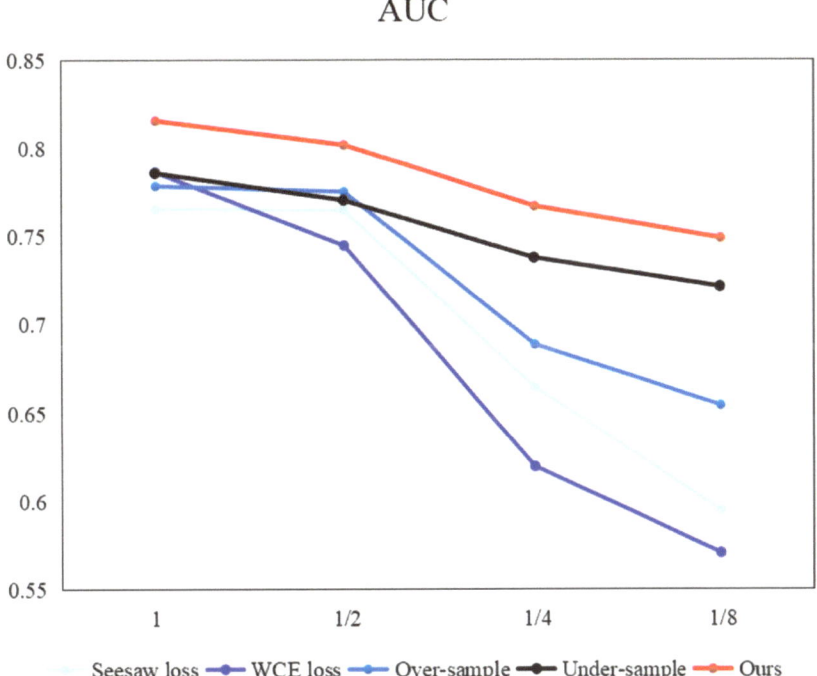

Figure 5. The AUC scores under different imbalance ratios.

4.3. The Influence on Cluster Number K

Cluster number K is an important parameter for the k-means++ algorithm. Different K will lead to a different cluster result. We set a series of K (3, 5, 7) and observe the impact on the result in Figure 6. Experiments show that our method is not sensitive to the choice of cluster number K. Different K can achieve similar performance. Although the number of clusters is different, our method has the same sampling proportion in different sub-groups, which ensures the distribution consistency between each balanced dataset and the original dataset. This mechanism makes our method more robust on hyper-parameter, that is, the cost of applying this method is low.

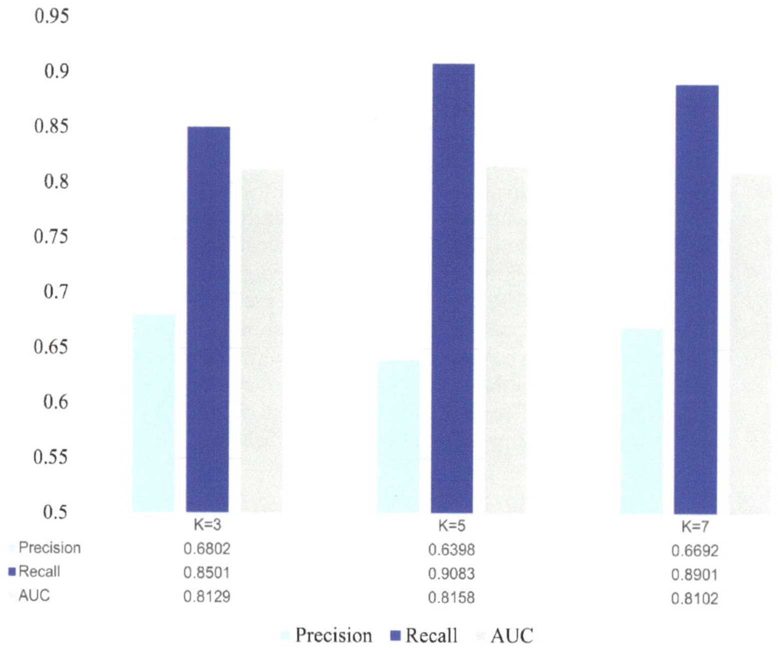

	K=3	K=5	K=7
Precision	0.6802	0.6398	0.6692
Recall	0.8501	0.9083	0.8901
AUC	0.8129	0.8158	0.8102

Figure 6. The results of the proposed method under different cluster number.

5. Conclusions

Data imbalance is very common in casting defect recognition. This paper proposes a distribution-preserving under-sampling method for reducing the uncertainty of traditional re-sampling methods. Our method divides all normal samples into several sub-groups by k-means++ and reassembles them into some balance datasets, which makes the normal samples in all balance datasets have an identical distribution with the original imbalance dataset. Experiments on our dataset illustrate that the proposed method achieves significant advantages compared to the state-of-the-art methods for data imbalance. At the same time, our method is more resistant to extreme data imbalance and not sensitive to the hyper-parameter.

Author Contributions: Conceptualization, H.Y.; methodology, H.Y.; software, H.Y.; validation, X.L. (Xinyue Li); formal analysis, X.L. (Xinyue Li); investigation, X.L. (Xinyue Li); resources, S.L.; data curation, C.H.; writing—original draft preparation, X.L. (Xinyue Li); writing—review and editing, H.X.; visualization, H.X.; supervision, H.X.; project administration, H.X.; funding acquisition, X.L. (Xingjie Li). All authors have read and agreed to the published version of the manuscript.

Funding: This research and APC were funded by CENTRAL-GUIDED FUND OF LOCAL DEVELOPMENT IN SCIENCE AND TECHNOLOGY, grant number 2022JH6/100100011.

Institutional Review Board Statement: Not applicable.

Informed Consent Statement: Not applicable.

Data Availability Statement: Not applicable.

Acknowledgments: The authors thank the anonymous reviewers and copy editor for valuable comments proposed.

Conflicts of Interest: The authors declare no conflict of interest.

References

1. Mery, D.; Filbert, D. Automated flaw detection in aluminum castings based on the tracking of potential defects in a radioscopic image sequence. *IEEE Trans. Robot. Autom.* **2002**, *18*, 890–901. [CrossRef]
2. Hernández, S.; Sáez, D.; Mery, D.; Silva, R.D.; Sequeira, M. Automated defect detection in aluminum castings and welds using neuro-fuzzy classifiers. In Proceedings of the 16th World Conference on NDT, Montreal, QC, Canada, 30 August 2004.
3. Zhao, X.; He, Z.; Zhang, S.; Liang, D. A sparse-representation-based robust inspection system for hidden defects classification in casting components. *Neurocomputing* **2015**, *153*, 1–10. [CrossRef]
4. Du, W.; Shen, H.; Fu, J.; Zhang, G.; He, Q. Approaches for improvement of the X-ray image defect detection of automobile casting aluminum parts based on deep learning. *NDT E Int.* **2019**, *107*, 102144. [CrossRef]
5. Ren, S.; He, K.; Girshick, R.; Sun, J. Faster R-CNN: Towards Real-Time Object Detection with Region Proposal Networks. *IEEE Trans. Pattern Anal. Mach. Intell.* **2015**, *39*, 1137–1149. [CrossRef] [PubMed]
6. Lin, T.Y.; Dollár, P.; Girshick, R.; He, K.; Hariharan, B.; Belongie, S. Feature Pyramid Networks for Object Detection. In Proceedings of the IEEE Conference on Computer Vision and Pattern Recognition, Honolulu, HI, USA, 21 July 2017.
7. Yu, H.; Li, X.; Song, K.; Shang, E.; Liu, H.; Yan, Y. Adaptive depth and receptive field selection network for defect semantic segmentation on castings X-rays. *NDT E Int.* **2020**, *116*, 102345. [CrossRef]
8. Liu, J.; Kim, J.H. A Variable Attention Nested UNet++ Network-Based NDT X-ray Image Defect Segmentation Method. *Coatings* **2022**, *12*, 634. [CrossRef]
9. Mery, D.; Arteta, C. Automatic Defect Recognition in X-Ray Testing Using Computer Vision. In Proceedings of the IEEE Winter Conference on Applications of Computer Vision, Santa Rosa, CA, USA, 27 March 2017.
10. Mery, D.; Riffo, V.; Zscherpel, U. GDXray: The Database of X-ray Images for Nondestructive Testing. *J. Nondestruct. Eval.* **2015**, *34*, 1–12. [CrossRef]
11. Mery, D. Aluminum Casting Inspection Using Deep Learning: A Method Based on Convolutional Neural Networks. *J. Nondestruct. Eval.* **2020**, *39*, 1–12. [CrossRef]
12. Tang, Z.; Tian, E.; Wang, Y.; Wang, L.; Yang, T. Non-destructive defect detection in castings by using spatial attention bilinear convolutional neural network. *IEEE Trans. Ind. Inform.* **2020**, *17*, 82–89. [CrossRef]
13. Hu, C.; Wang, Y. An efficient cnn model based on object-level attention mechanism for casting defects detection on radiography images. *IEEE Trans. Ind. Electron.* **2020**, *67*, 10922–10930. [CrossRef]
14. Jiang, L.; Wang, Y.; Tang, Z.; Miao, Y.; Chen, S. Casting defect detection in X-ray images using convolutional neural networks and attention-guided data augmentation. *Measurement* **2021**, *170*, 108736. [CrossRef]
15. Wang, J.; Zhang, W.; Zang, Y.; Cao, Y.; Pang, J.; Gong, T.; Chen, K.; Liu, Z.; Loy, C.C.; Lin, D. Seesaw loss for long-tailed instance segmentation. In Proceedings of the IEEE/CVF Conference on Computer Vision and Pattern Recognition, Nashville, TN, USA, 19 June 2021.
16. Lin, T.Y.; Goyal, P.; Girshick, R.; He, K.; Dollár, P. Focal loss for dense object detection. In Proceedings of the IEEE International Conference on Computer Vision, Venice, Italy, 22 October 2017.
17. Estabrooks, A.; Jo, T.; Japkowicz, N. A multiple resampling method for learning from imbalanced data sets. *Comput. Intell.* **2004**, *20*, 8–36. [CrossRef]
18. Liu, X.-Y.; Wu, J.; Zhou, Z.-H. Exploratory undersampling for class-imbalance learning. *IEEE Trans. Syst. Man Cybern.* **2008**, *39*, 539–550.
19. Arthur, D.; Vassilvitskii, S. K-means++: The Advantages of Careful Seeding. In Proceedings of the Eighteenth Annual ACM-SIAM Symposium on Discrete Algorithms, Miami, FL, USA, 22 January 2006.
20. He, K.; Zhang, K.; Ren, S.; Sun, J. Deep residual learning for image recognition. In Proceedings of the IEEE Computer Society Conference on Computer Vision and Pattern Recognition, Las Vegas, NV, USA, 16 June 2016.
21. Deng, J.; Dong, W.; Socher, R.; Li, L.J.; Li, K.; Fei-Fei, L. ImageNet: A Large-Scale Hierarchical Image Database. In Proceedings of the IEEE Computer Society Conference on Computer Vision and Pattern Recognition, Miami, FL, USA, 18 June 2009.
22. Sandler, M.; Howard, A.; Zhu, M.; Zhmoginov, A.; Chen, L.C. MobileNetV2: Inverted Residuals and Linear Bottlenecks. In Proceedings of the IEEE/CVF Conference on Computer Vision and Pattern Recognition, Salt Lake City, UT, USA, 17 June 2018.

Article

Multi-Scale Lightweight Neural Network for Steel Surface Defect Detection

Yichuan Shao [1], Shuo Fan [2], Haijing Sun [1], Zhenyu Tan [2], Ying Cai [2], Can Zhang [2] and Le Zhang [1,*]

[1] School of Intelligent Science & Engineering, Shenyang University, Shenyang 110044, China; shaoyichuan@syu.edu.cn (Y.S.); suhaijing@syu.edu.cn (H.S.)

[2] School of Information Engineering, Shenyang University, Shenyang 110044, China; fanshuoi2022@163.com (S.F.); tanzhenyu@syu.edu.cn (Z.T.); caiying@syu.edu.cn (Y.C.); zhangcan@syu.edu.cn (C.Z.)

* Correspondence: zhangle@syu.edu.cn

Abstract: Defect classification is an important aspect of steel surface defect detection. Traditional approaches for steel surface defect classification employ convolutional neural networks (CNNs) to improve accuracy, typically by increasing network depth and parameter count. However, this approach overlooks the significant memory overhead of large models, and the incremental gains in accuracy diminish as the number of parameters increases. To address these issues, a multi-scale lightweight neural network model (MM) is proposed. The MM model, with a fusion encoding module as its core, constructs a multi-scale neural network by utilizing the Gaussian difference pyramid. This approach enhances the network's ability to capture patterns at different resolutions while achieving superior model accuracy and efficiency. Experimental results on a dataset from a hot-rolled strip steel plant demonstrate that the MM network achieves a classification accuracy of 98.06% in defect classification tasks. Compared to networks such as ResNet-50, ResNet-101, VGG, AlexNet, MobileNetV2, and MobileNetV3, the MM model not only reduces the number of model parameters and compresses model size but also achieves better classification accuracy.

Keywords: surface defect detection; defect classification; deep learning; lightweight network

1. Introduction

Defect classification is an important industrial inspection task where defect images are analyzed and identified to determine their corresponding defect types [1]. Manual defect detection is commonly used to classify steel surface defects in traditional industries. It is an essential component of the industrial defect detection process. To replace manual operations, it is desired that machines can automatically detect steel surface defects using computer vision technology [2].

The task of classifying defects on steel surfaces using computer vision techniques poses a significant challenge due to the effects of illumination and material variations on defect images [3]. In addition, the appearance of defects varies dramatically not only within categories of steel surfaces but also between categories, thus further complicating the classification process. Therefore, designing an accurate and reliable defect classification algorithm to take these complexities into account is an ongoing research topic in the field of computer vision [4]. Current image classification methods are mainly of two types: traditional machine learning image classification algorithms and deep learning methods based on convolutional neural networks [5]. Traditional image classification algorithms are mainly implemented using two major steps of feature extraction and classifier design, such as the K-Nearest Neighbor algorithm [6], Support Vector Machine [7], and neural networks [8], which are widely used in computer vision. Various complex situations are faced in practical defect classification applications, and it is difficult to achieve the requirements in terms of accuracy using traditional image processing methods [9].

The identification of surface defects in steel undergoes three processes: manual human detection, prediction using machine learning algorithms, and automatic detection through deep learning [10]. In recent years, deep learning-based image classification methods have achieved good results, such as VGGNet (Visual Geometry Group Network) [11] and ResNet (Residual Network) [12]. U-Net is employed for the task of detecting and classifying welding defects using X-ray images, artificial neural networks, and image analysis methods [13,14]. We introduced a novel deep learning-based approach for detecting and classifying surface defects that occur during the steel production process. This method enhances classification performance through parallel training of residual and attention structures. LSHADE-SVC-PCD proposes an intelligent method for automatic detection of pitting corrosion by employing LSHADE metaheuristics, SVM machine learning, and image processing techniques [15,16]. We proposed a defect detection algorithm based on deformable networks combined with a multi-scale feature fusion algorithm leveraging a deformable convolutional neural network. However, state-of-the-art CNNs require billions of floating-point operations, which makes them unusable for mobile or embedded devices. For example, ResNet-101 has a complexity of 7.8×10^9 FLOPs (floating-point operations per second), which makes real-time detection impossible even with powerful GPUs. Considering the huge computational cost of modern CNNs, lightweight neural networks have been proposed to be deployed on mobile or embedded devices. For example, MobileNetV1 [17] and MobileNetV2 [18] use depth-separable convolution to build lightweight networks. ShuffleNet [19] uses grouped convolution and depth-separable convolution to build lightweight networks. SqueezeNet [20] uses the core module Fire to compress the model parameters, reduce the depth of the network, and decrease the size of the model. SENet (Squeeze-and-Excitation Network) [21] proposes the SE module as a lightweight attention mechanism to adaptively calibrate the feature map by learning the channel importance; however, the SE module only focuses on the influence of the channel aspect of the feature map and ignores the importance of the spatial dimension.

Lightweight networks [22] can achieve relatively high accuracy with limited computational budgets. However, existing lightweight networks tend to use "sparsely connected" convolution, such as deep convolution and group convolution, rather than the standard "fully connected" convolution. This "sparsely connected" convolution, while reducing the number of parameters, can to some extent hinder the exchange of information between groups, resulting in a degradation of network performance. Since a practical steel defect classification algorithm needs to be deployed on CPUs or even embedded systems, an algorithm with low computational complexity and high classification accuracy that can avoid intergroup information loss is needed.

In this paper, a novel multiscale neural network model (MM) is proposed to build a multiscale neural network model with a fusion coding module as the core and a Gaussian difference pyramid to obtain better model accuracy and efficiency while improving the network's ability to capture patterns of different resolutions. The feasibility and effectiveness of the model and method are verified through experiments on the surface defect detection of steel sheet products in a factory. By comparing the results with other methods, it can be concluded that the multiscale neural network model avoids the loss of information between groups, further reduces the number of parameters and computational effort, and significantly improves classification accuracy.

Our main contributions can be summarized as follows:

(1) We propose a method that utilizes the Gaussian difference pyramid to construct a scale space by iteratively building a pyramid structure at different scales and effectively detecting key points in images through a scale-invariant feature transform.
(2) We employ the multi-kernel fusion approach to capture both blurred and fine-grained features at different resolutions, enhancing the accuracy and efficiency of the model.
(3) By integrating the fusion encoding and matching transformation of keypoints with the original network, we address the challenges of exchanging information, avoiding information loss, and improving performance in multi-scale space.

(4) The proposed Multi-Scale Lightweight Neural Network (MM) achieves better classification accuracy while reducing model parameters and compressing model size.

The structure of this paper is as follows: Section 1 presents an overview of the proposed model's overall structure. Section 2 discusses the method employed to address the overfitting issue. In Section 3, the dataset and evaluation of the proposed model are presented in detail, followed by a discussion of the results. Finally, Section 4 concludes the paper.

2. Construction of a Multi-Scale Neural Network Model

The problem of compression and acceleration of neural networks has become a hot research topic in the field of deep learning due to the large computational volume and model capacity of deep neural networks. As the demand for high-quality deep neural networks running on embedded devices increases, researchers are exploring the design of lightweight network models to reduce computational costs. These models often use "sparsely connected" convolutions, which can lower computational demands but may also inhibit information exchange between different groups within the network. In this paper, we propose a multiscale neural network by borrowing the idea of the Gaussian difference pyramid [23] and adding a fusion coding module [24]. The network improves classification accuracy while decreasing the computational cost.

2.1. Constructing Key-Point Feature Sets Based on Scale Space

In the steel slice dataset, the representation of images is represented at multiple scales. To effectively capture these features at different scales, we propose a method to construct the scale space using Gaussian [25] difference pyramids. In the scale space construction module, we construct the scale space by Gaussian blurring the image and calculating the difference between adjacent Gaussian blurred images. This process is iterated over different scales (i.e., the size or resolution of the image), forming a kind of pyramid structure. In this structure, each layer of the image is more blurred compared to the previous layer, and the resolution is reduced accordingly. By using this scale-space building block, we are able to effectively detect key points in the image during the scale-invariant feature transform (SIFT).

Gaussian differential pyramids provide an effective scale-space representation that captures the features of an image at different scales [26]. This is important for dealing with real-world image problems, as real-world objects can appear at a variety of different scales. In addition, the Gaussian difference pyramid offers the advantage of precise localization of feature points in images and provides multi-scale image information for analysis and processing. Furthermore, the downsampling of images in the pyramid reduces computational complexity, thereby enhancing computational efficiency. These characteristics make the Gaussian difference pyramid widely utilized in the fields of image processing, computer vision, and pattern recognition.

2.2. Convolution of Key-Point Features

The traditional convolutional layer in convolutional neural networks (CNNs) preserves global spatial information during the convolution process but may overlook crucial details. Due to the lack of explicit focus on key points, traditional convolutional layers may not be as effective as key-point feature convolution in extracting important features. Additionally, traditional convolutional layers do not specifically handle key points, leading to a lower sensitivity for accurately capturing significant key points in images. This limitation can restrict the performance of models on tasks related to key points. Key-point feature convolution is a group convolution of each key-point feature set into a separate set, which reduces the number of parameters and computational cost [27]. Using the idea of multi-core combination, convolution using different key point feature sets can enhance the model's ability to adapt to varying levels of detail. The network needs both fuzzy key-point features to capture high-resolution patterns and fine key-point features to capture low-resolution patterns for better model accuracy and efficiency.

2.3. Key-Point Feature Set Mapping Fusion Module

The key-point feature set mapping fusion module has an advantage over traditional convolutional layers in convolutional neural networks (CNNs) as it explicitly focuses on key points and integrates their important information into the network, thereby enhancing the perception and utilization of crucial details in images and improving the robustness and accuracy of the model. On the other hand, traditional convolutional layers exhibit drawbacks in handling key details and tasks related to key points, such as disregarding fine details, a limited receptive field, and a lack of explicit attention to key points. To address issues related to information exchange, loss, and performance degradation of key-point features in multi-scale space [28], The paper suggests using key points to map and encode information from different scales, then combining it with the original network through matching transformations in order to enhance the discriminative features.

Key-point feature mapping fusion aims to combine essential feature points and incorporate information about their varying scales in order to create a comprehensive feature map, the process is shown in Figure 1. On top of the original feature map F ∈ RC × H × W generated by the Gaussian difference pyramid, the fusion module performs the fusion encoding transformation TF: RC × H × W → RCM × H × W to achieve the purpose of aggregating all key-point features. Where: C is the key-point feature of the original feature map; H and W are the width and height of the original feature map; CM is the key-point feature of the fused feature map.

The image is decomposed by a Gaussian differential pyramid to produce N different resolution images, and the Gaussian differential pyramid consists of multiple groups of pyramids, where each group of pyramids contains several layers, and the Gaussian differential pyramid consists of layer orders constructed on the basis of the Gaussian pyramid [29]. The process of decomposing a Gaussian image into its differential pyramid involves:

Step 1. Initialize $i = 0$;
Step 2. Standard image $I(x,y)$ is sampled to obtain the first layer of the first set of Gaussian pyramid images $g_{0,0}$;
Step 3. Initialize $j = 0$ and $x = 0$;
Step 4. The Gaussian kernel Gx is convolved with image $g_{i,0}$ [30]:

$$G_x(x, y, \sigma_x) = \frac{1}{2\pi\sigma_x^2} e^{\frac{(x-x_0)^2 + (y-y_0)^2}{2\sigma_x^2}} \quad (1)$$

$$g_{i,j+1}(x, y) = g_{i,j}(x, y) \otimes G_x(x, y, \sigma_x) \quad (2)$$

where σ_x is the smoothing parameter.
Step 5. Differentiate the Gaussian image from the Gaussian image to obtain the Gaussian difference image [31]:

$$d_{i,x}(x, y) = g_{i,j}(x, y) - g_{i,j+1}(x, y) \quad (3)$$

Step 6. $j = j + 1$ and $x = x + 1$, iterate Steps 4 and 5, and when $j > n - 1$ and $x > n - 2$, perform Step 7;
Step 7. Downsample the image to get the Gaussian image of layer $i + 1$. When $i = I + 1$, go to Step 3, and the decomposition process ends when $i > m - 1$ is satisfied.

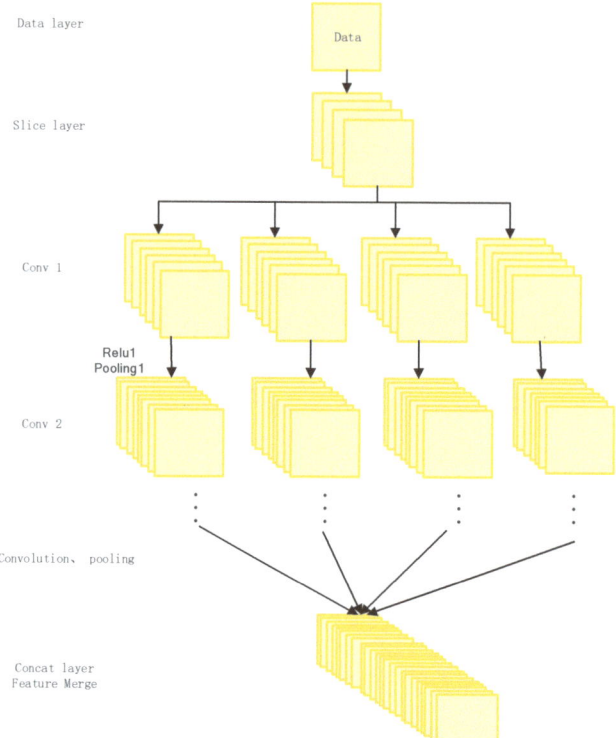

Figure 1. Key-point feature convolution and mapping fusion.

2.4. MM Network Construction

Based on Section 2.1 "Constructing Key-Point Feature Sets Based on Scale Space", Section 2.2 "Convolution of Key-Point Features" and Section 2.3 "Key-Point Feature Set Mapping Fusion Module", the MM network is proposed in this paper. The MM network is shown in Figure 2. The network consists of three branches from bottom to top: the standard branch, the dimension reduction branch, and the fusion branch. The standard branch directly maps the original feature map. The dimension reduction branch reduces computational costs and then performs Section 2.1 "Constructing Key-Point Feature Sets Based on Scale Space" operations to construct a set of key-point features. The fusion branch utilizes mixed convolutional kernels to obtain more stable feature maps at different resolutions, where the large convolutional kernel in the mixed convolution retains more feature information. It serves as a bridge connecting the dimension reduction branch and the fusion branch by combining the processed feature maps from the fusion branch. The final single-point group convolution is used to restore the channel dimension to match the standard branch. The fusion branch performs fusion encoding on the network and combines with the dimension reduction branch using element-wise multiplication before the single-point group convolution, which helps to reduce the loss of inter-group information during the convolution process. In this figure, "GC" represents Group Convolution, "Conv" represents Convolution, and "BN" represents Batch Normalization.

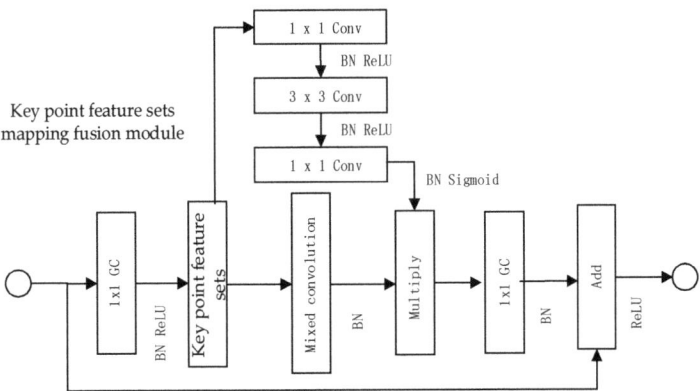

Figure 2. MM network construction.

3. Solving the Overfitting Problem

To avoid overfitting problems, We perform batch normalization of the network as well as Dropout operations on the last two fully connected layers of the network [32]. Unlike L1 and L2 normalization, Dropout does not rely on the modification of the cost function; in Dropout, the network itself is changed. Assuming training data x and a corresponding target output y, the contribution to the gradient is normally determined by forward propagating x in the network and then back-propagating. However, using the Dropout operation, this paper starts by randomly removing half of the hidden neurons in the network while leaving the neurons in the input and output layers unchanged, after which the input x is forward propagated and the contribution to the gradient is determined by modifying the network, back-propagating the result, and repeating the process. The overall execution process is: first reset the neurons of Dropout; then select a new random subset of hidden neurons for deletion; and finally, update the weights and biases by estimating the gradient for a different small batch of data before determining.

Batch normalization is a normalization technique applied in hidden layers that normalizes the inputs of each layer by subtracting the batch mean and dividing by the batch standard deviation. This normalization process helps stabilize the distribution of activation values in the network, reducing the range of gradient variations, thereby accelerating training and improving the generalization ability of the network.

In practice, batch normalization not only improves training effectiveness but also has a certain regularization effect. By introducing randomness in each batch, it reduces reliance on dropouts. This is because the introduced randomness in batch normalization reduces the dependence of the network on individual neurons, thus reducing the risk of overfitting. Therefore, in networks with batch normalization, lower dropout rates can typically be used, or dropout operations may not be necessary at all.

Batch normalization plays a significant role in reducing dropout rates. By standardizing the inputs and stabilizing the distribution of activation values in the network, the reliance on Dropout operations is alleviated. This allows for a better balance between model capacity and the risk of overfitting, thereby improving the generalization ability and training effectiveness of the model.

4. Experiment

This section first describes the system that utilizes a CCD camera to capture images of the surface of hot-rolled steel strips, which are then subjected to normalization preprocessing. A sliding window approach is employed to extract image patches, which are used to construct a dataset of standard defect images. The dataset consists of six defect categories, including transverse cracks, wrinkles, longitudinal cracks, edge cracks, seams, and water stains. The training and validation of the system are performed using a multi-scale lightweight neural network model, and the performance of the model is evaluated and analyzed. Experimental results demonstrate that the proposed model achieves high accuracy and efficiency in detecting surface defects in hot-rolled steel strips. Compared to other classical and lightweight neural network models, the proposed model exhibits superior capability in detecting minor defects with high classification accuracy and performance.

4.1. Common Defects and Detection Process

The defect images are acquired based on a machine vision technology acquisition system, using a system consisting of CCD cameras as well as a deep neural network server. The images of the hot rolled strip surface are obtained through CCD cameras, and then the image data are received and transmitted to process the defect images of the hot rolled strip surface obtained within the cameras. The defect images are acquired based on a machine vision technology acquisition system, using a system consisting of CCD cameras as well as a deep neural network server. The images of the hot rolled strip surface are obtained through CCD cameras, and then the image data are received and transmitted to process the defect images of the hot rolled strip surface obtained within the cameras [33]. The data sets used in this paper are obtained through such image acquisition units, taken from a hot-rolled strip mill specializing in the production of cover hot-rolled strip surfaces, with more than 20,000 original sample images taken from different production batches.

There are gaps in the images of hot-rolled strip steel surface defects acquired under different light intensities, light directions, etc. Therefore, the original images are normalized and pre-processed. In this paper, a sliding window of 128×128 pixels is used to intercept the whole CCD-captured hot-rolled strip surface defect images, and the complete defect as well as defect-free images are selected to establish a standard image dataset of hot-rolled strip surface defects, which reaches more than 60,000 standard sample images. This paper artificially augments the standard image dataset by adding rotated images to enhance model training quality, reaching a total of more than 120,000 standard sample images, which contain horizontal cracking, pleats, Side Splits, seams, vertical cracks, and water stains (as seen in Figure 3).

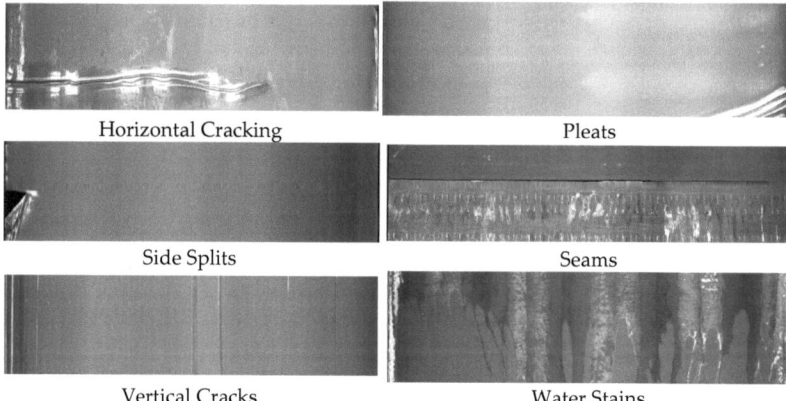

Figure 3. Example defect image and corresponding label.

Among them, horizontal cracking has the greatest impact on the quality of hot-rolled strip and is the biggest safety hazard, while the effects of folds, pleats, side splits, seams, vertical cracks, and water stains decrease in order. The obtained standard sample image dataset was divided into 3 parts: partitioning the data into subsets based on a specific proportion (as shown in Table 1).

Table 1. Standard defect image data set.

Dataset	Horizontal Cracking	Pleats	Vertical Cracks	Side Split	Seams	Water Stains	Total
Training set	20,000	20,000	15,000	15,000	15,000	15,000	100,000
Validation set	2000	2000	1500	1500	1500	1500	10,000
Test set	2000	2000	1500	1500	1500	1500	10,000

The dataset faces three main challenges: (1) defects within classes vary greatly in appearance; (2) defects between classes have similar aspects, and (3) the grayscale of defect images between classes can change due to the effect of defect images on illumination and material variations.

For the content and criteria of common defects on the surface of a hot-rolled strip of cover, this paper uses the constructed multi-scale convolutional neural network model to experimentally validate the detection of defects on the surface of a hot-rolled strip, and the main process is as follows:

Step 1. Selection of the Pytorch open-source framework for deep learning as the experimental environment for building multi-scale convolutional neural network models.

Step 2. While constructing the experimental environment, the CCD camera acquires image data of the surface of the hot rolled strip, and the images of the defects on the surface of the hot rolled strip obtained within the camera are processed and displayed on the monitor.

Step 3. Intercept the images of defects on the surface of the hot-rolled strip obtained in the camera, select the complete defective and defect-free images, normalize the detection of "defective images", obtain the image size of 128 × 128, establish the defective standard image data set, and form the experimental sample.

Step 4. Training and validation of the samples based on a multiscale convolutional neural network model labeled with six types of defects.

Step 5. Randomly selected images from each category in the standard sample image dataset become the test samples, and the classification results and accuracy of the model to detect surface defects in images of hot-rolled strip steel are analyzed.

The whole process is shown in Figure 4.

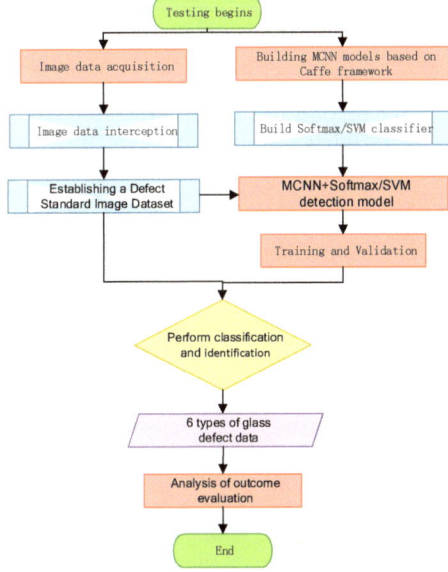

Figure 4. Detection process.

4.2. Model Training

The MM models experimentally designed in this paper are performed based on the open-source framework of Pytorch deep learning, which provides a complete toolkit for training, testing, fine-tuning, and developing models. The models and corresponding optimizations are given in textual form rather than in code.

Steps during model training tests:

Step 1. Interception of the entire CCD-captured image of the hot-rolled strip surface defects using a sliding window of 128×128 pixels, selection of the complete defects, image normalization to obtain a standard defect picture, and collation into a standard defect data set.

Step 2. SIFT extraction of Gaussian differential pyramidal images from standard defect images to obtain a dataset for training a multi-scale convolutional neural network.

Step 3. The obtained multi-resolution image training data set is directly input to the network, and the four multi-scale images are divided by the slice layer and convolved separately to extract features. Initialize the network weights with "Gaussian", where the bias is set to "Constant".

Step 4. Select a batch training sample from the training set and input it into the network.

Step 5. Samples are propagated forward through the mapping between layers until Concat for feature merging, and then continue to propagate forward until the output layer to obtain the actual output vector.

Step 6. Calculates the error between the actual output vector and the label, and if the error is less than a predetermined threshold (or if the number of training iterations reaches a predetermined threshold), the network training is stopped; otherwise, the network continues.

Step 7. Tuning the weight parameters of the whole network model by backpropagation according to the principle of minimum error cost.

Step 8. Revert to Step 4 and continue the training.

Step 9. A randomly selected test dataset is fed into the trained model (convolutional kernel set, network weight parameters, etc.) for recognition detection.

The multi-scale defect image detection models constructed in this paper are all supervised training methods, and their training image sets are composed of vector pairs of (defect image, category label), where "defect image" is a normalized image, and the size of the image obtained after normalization is 128×128. The "category labels" represent the classification labels of the input defect images, which are divided into six categories: horizontal cracks, folds, vertical cracks, edge cracks, seams, and water stains.

In training, the experiments were conducted on the Ubuntu 20.04.2 LTS operating system, using the PyTorch deep learning framework and Python as the programming language. The CPU utilized was an Intel Core i7-9700F, while the GPU employed was an NVIDIA GeForce RTX 2080Ti. The images in the input layer are fixed-size 128×128 RGB images, and for the training set, the preprocessing is performed to subtract the average RGB value per pixel. The training network parameters are set as follows: The network uses the AdamW [34] optimization algorithm in the training phase to iteratively update the weights of the neural network based on the training data and the selected small batch training size; each batch contains 64 images. The weight decay is 0.005, the memory factor is 0.9, the learning rate is 0.001, and the learning strategy is STEP. The normalization factor is used to accelerate the training process in GPU mode, and the maximum number of training iterations is 1000.

4.3. Analysis of Results

The Figure 5 illustrates the training process of the MM model network, including the curves for accuracy, training loss, and validation loss. It can be observed from the graph that with the increase in the number of iterations, the model's accuracy on the validation set continues to rise while its loss decreases continuously. The model exhibits good convergence and achieves an accuracy of around 98.4% after 700 iterations.

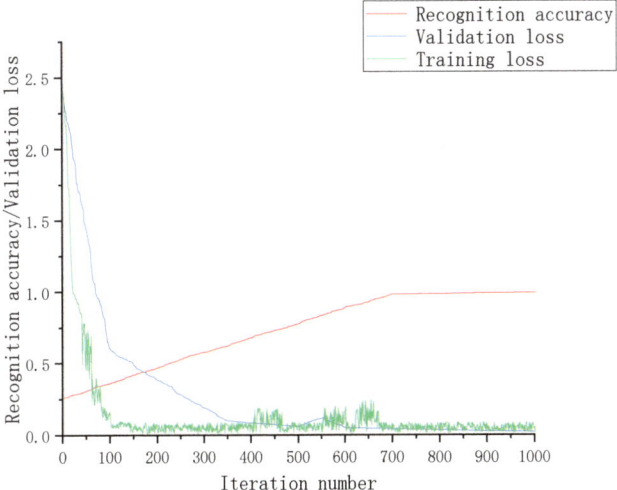

Figure 5. Defective image recognition accuracy and loss curves.

To evaluate the performance of our classification model, we have introduced a confusion matrix, which serves as a valuable tool for assessing prediction accuracy. The confusion matrix is a square matrix that represents the relationship between actual and predicted classes. Rows of the matrix correspond to the true classes, while columns denote the predicted classes. Each element in the matrix represents the count of instances classified into a specific combination of true and predicted classes. As shown in Table 2.

Table 2. Confusion matrices.

	Prediction of Horizontal Cracking	Prediction of Pleats	Prediction of Vertical Cracks	Prediction of Side Split	Prediction of Seams	Prediction of Water Stains
Actual Horizontal Cracking	1984	5	3	3	2	3
Actual Pleats	10	1987	1	1	0	1
Actual Vertical Cracks	6	5	1486	2	1	0
Actual Side Splits	4	4	1	1488	1	2
Actual Seams	5	2	1	2	1490	0
Actual Water Stains	2	0	0	1	3	1494

The first row represents the samples with the true label "horizontal cracking". Among these samples, 1984 were correctly predicted as "horizontal cracking", 5 were wrongly predicted as "pleats", 3 were wrongly predicted as "vertical cracks ", 3 were wrongly predicted as "side splits", 2 were wrongly predicted as "seams", and 3 were wrongly predicted as "water stains".

The second row represents the samples with the true label "pleats". Among these samples, 10 were wrongly predicted as "horizontal cracking", 1987 were correctly predicted as "pleats", 1 was wrongly predicted as "vertical cracks", 1 was wrongly predicted as "side splits", 0 were wrongly predicted as "seams", and 1 was wrongly predicted as "water stains".

The following rows follow the same pattern, representing the remaining labels. This confusion matrix helps us understand the performance of the MM model for each defect type and the possible misclassification of the model situation.

To verify the impact of different modules on network performance, five evaluation indices were introduced in this article: computing power, which refers to the number of floating-point operations executed per second (FLOPs), feature memory, recall, F1 score, and accuracy. The MM network was compared with four popular classical networks (ResNet-101, ResNet-50, VGG, and AlexNet) as well as the new lightweight neural networks MobileNetV2 and MobileNetV3. Table 3 shows the detailed comparison results.

Table 3. Comparison of the comprehensive performance of different networks.

Models	Precision	Feature Memory	Number of Floating-Point Operations (FLOPs)	Recall	F1
Resnet-50	89.7%	60 MB	10 GFLOPs	87.2%	0.92
ResNet-101	97.38%	155 MB	8 GFLOPs	92.6%	0.93
VGG	93.56%	96 MB	15.5 GFLOPs	93.7%	0.90
AlexNet	91.30%	300 MB	0.72 GFLOPs	91.6%	0.92
MobileNetV2	94.6%	57 MB	0.98 GFLOPs	94.6%	0.95
MobileNetV3	96.7%	52 MB	0.76 GFLOPs	95.4%	0.97
MM	98.06%	50 MB	0.67 GFLOPs	98.7%	0.99

After 1000 iterations, AlexNet, ResNet-50, and VGG achieved classification accuracies of 91.30%, 89.7%, and 93.56%, respectively, while ResNet-101, MobileNetV2, and MobileNetV3 achieved slightly better accuracies of 97.38%, 94.6%, and 96.7%, respectively. Compared to Resnet50, MM increased the accuracy by 8.36 percentage points, by 0.38 percentage points compared to ResNet-101, by 4.5 percentage points compared to VGG, by 6.76 percentage points compared to AlexNet, by 3.46 percentage points compared to MobileNetV2, and by 1.36 percentage points compared to MobileNetV3. Additionally, MM had a faster decrease in training loss compared to the other models, with a final loss value approaching zero as shown in Figure 6. Moreover, MM had smaller feature memory and fewer floating-point computations, while its recall and F1 metrics were relatively higher. These results fully demonstrate that the key-point feature convolutional operation reduces the number of network parameters and that the key-point feature set mapping fusion operation promotes information exchange between key points, making it effective in reducing performance losses.

From the comprehensive analysis of the experiments above, MM has the best performance in all indicators. Resnet-50, ResNet-101, VGG, AlexNet, MobileNetV2, and MobileNetV3 models lack the ability to detect small defects on the surface of steel. The MM proposed in this paper, with the fusion encoding module as the core, constructs a multiscale neural network model through the Gaussian difference pyramid. It not only improves the network's capture ability for different resolution modes but also achieves better model accuracy and efficiency. The recognition accuracy on the steel surface defect dataset is the best. The recognition accuracy of various neural network models for different types of defects is shown in Table 4.

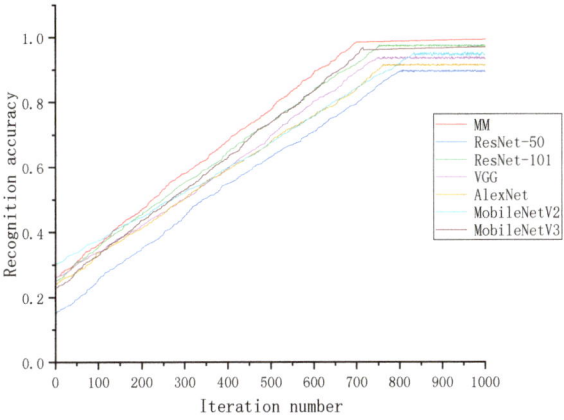

(a) Verification of the accuracy curve.

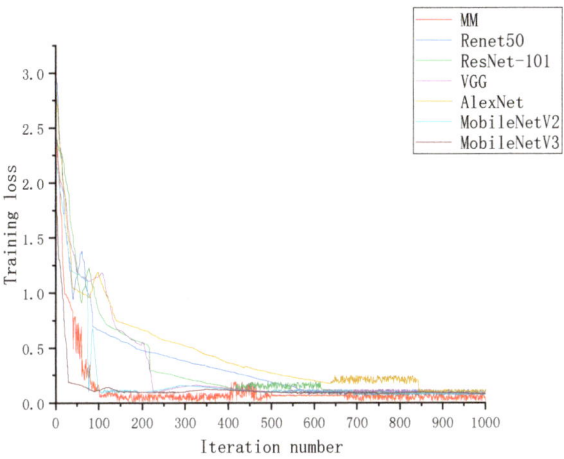

(b) Training loss curves

Figure 6. Performance comparison of different models.

Table 4. Accuracy of different networks for defect recognition in various categories.

Transverse Crack	ResNet-50	ResNet-101	VGG	AlexNet	MobileNetV2	MobileNetV3	MM
Horizontal Cracking	92.4%	91.6%	86.2%	92.1%	91.2%	92.6%	99.2%
Pleats	72.6%	98.2%	82.9%	97.2%	97.7%	97.2%	99.3%
Side Splits	93.2%	97.2%	92.6%	92.2%	98.6%	98.6%	99.2%
Seams	81.7%	98.9%	93.5%	93.2%	95.8%	93.5%	99.3%
Water Stains	91.2%	99.7%	92.4%	95.2%	96.6%	99.2%	99.6%
Vertical cracks	93.9%	93.2%	97.9%	96.2%	94.2%	98.7%	99.0%

5. Conclusions

To address the issues of increasing parameter count and computational cost in convolutional neural networks as well as the challenge of preserving key-point feature information in existing lightweight networks, this paper proposes a novel lightweight network, referred to as MM. The core of this model is a fusion encoding module that leverages Gaussian difference pyramids to construct a multi-scale neural network model. By enhancing the network's ability to capture patterns at different resolutions, the proposed model achieves higher model accuracy and efficiency. Experimental results show that the network avoids the loss of feature information at key points, improves the classification accuracy, and significantly improves the overall performance compared with other networks, which provides strong support for the mobile classification task of steel surface defects. It provides strong support for the mobile deployment of steel surface defect classification tasks. In the future, we plan to improve the model's ability to perform well on new data and optimize it to meet the requirements for commercialization. Furthermore, the proposed model can be applied to real-time object detection and tracking tasks, mobile device applications, and image super-resolution reconstruction. These directions leverage the lightweight nature and multi-scale capabilities of MM, offering possibilities for real-time image analysis, edge computing, and enhanced image processing in various domains.

Author Contributions: Conceptualization, Y.S.; Methodology, Y.S., S.F. and H.S.; Software, S.F., Z.T., Y.C. and C.Z.; Validation, C.Z. and L.Z.; Formal analysis, L.Z.; Investigation, L.Z.; Resources, Z.T.; Data curation, S.F.; Writing—original draft, H.S., Z.T. and L.Z. All authors have read and agreed to the published version of the manuscript.

Funding: This research was funded by Shenyang Science and Technology Plan (grant number 22-319-2-26) and the APC was funded by Shenyang University.

Institutional Review Board Statement: Not applicable.

Informed Consent Statement: Not applicable.

Data Availability Statement: Not applicable.

Conflicts of Interest: The authors declare no conflict of interest.

References

1. He, Y.; Wen, X.; Xu, J. A Semi-Supervised Inspection Approach of Textured Surface Defects under Limited Labeled Samples. *Coatings* **2022**, *12*, 1707. [CrossRef]
2. He, Y.; Song, K.; Meng, Q.; Yan, Y. An End-to-End Steel Surface Defect Detection Approach via Fusing Multiple Hierarchical Features. *IEEE Trans. Instrum. Meas.* **2020**, *69*, 1493–1504. [CrossRef]
3. Liu, Y.; Yuan, Y.; Balta, C.; Liu, J. A Light-Weight Deep-Learning Model with Multi-Scale Features for Steel Surface Defect Classification. *Materials* **2020**, *13*, 4629. [CrossRef] [PubMed]
4. Boudiaf, A.; Benlahmidi, S.; Harrar, K.; Zaghdoudi, R. Classification of Surface Defects on Steel Strip Images using Convolution Neural Network and Support Vector Machine. *J. Fail. Anal. Prev.* **2022**, *22*, 531–541. [CrossRef]
5. Nizan, O.; Tal, A. k-NNN: Nearest Neighbors of Neighbors for Anomaly Detection. *arXiv* **2023**, arXiv:2305.17695. [CrossRef]
6. Shamsi, M.; Beheshti, S. Separability and Scatteredness (S&S) Ratio-Based Efficient SVM Regularization Parameter, Kernel, and Kernel Parameter Selection. *arXiv* **2023**, arXiv:2305.10219. [CrossRef]
7. Melgani, F.; Bruzzone, L. Classification of hyperspectral remote sensing images with support vector machines. *IEEE Trans. Geosci. Remote Sens.* **2004**, *42*, 1778–1790. [CrossRef]
8. Wen, X.; Shan, J.; He, Y.; Song, K. Steel Surface Defect Recognition: A Survey. *Coatings* **2023**, *13*, 17. [CrossRef]
9. Yang, Z.; Zhang, M.; Chen, Y.; Ping, E. Research progress on surface defect detection methods based on machine vision. *Mod. Manuf. Eng.* **2023**, *511*, 143. [CrossRef]
10. Choi, E.; Schuetz, A.; Stewart, W.F.; Sun, J. Using Recurrent Neural Network Models for Early Detection of Heart Failure Onset. *J. Am. Med. Inform. Assoc.* **2017**, *24*, 361–370. Available online: https://academic.oup.com/jamia/article/24/2/361/2631499?login=false (accessed on 23 June 2023). [CrossRef]
11. Simonyan, K.; Zisserman, A. Very Deep Convolutional Networks for Large-Scale Image Recognition. *arXiv* **2015**, arXiv:1409.1556. [CrossRef]
12. He, K.; Zhang, X.; Ren, S.; Sun, J. Deep Residual Learning for Image Recognition. In Proceedings of the IEEE Conference on Computer Vision and Pattern Recognition, Las Vegas, NV, USA, 27–30 June 2016; pp. 770–778. Available online: https://openaccess.thecvf.com/content_cvpr_2016/html/He_Deep_Residual_Learning_CVPR_2016_paper.html (accessed on 23 June 2023).
13. Kothari, J.D. Detecting Welding Defects in Steel Plates using Machine Learning and Computer Vision Algorithms. *Int. J. Adv. Res. Electr. Electron. Instrum. Eng.* **2018**, *7*, 3682–3686. Available online: https://papers.ssrn.com/abstract=3729754 (accessed on 23 June 2023).

14. Demir, K.; Ay, M.; Cavas, M.; Fatih, D. Automated Steel Surface Defect Detection and Classification Using a New Deep Learning-Based Approach. *Neural Comput. Appl.* **2023**, *35*, 8389–8406. Available online: https://link.springer.com/article/10.1007/s00521-022-08112-5 (accessed on 23 June 2023). [CrossRef]
15. Hoang, N.-D. Image Processing-Based Pitting Corrosion Detection Using Metaheuristic Optimized Multilevel Image Thresholding and Machine-Learning Approaches. *Math. Probl. Eng.* **2020**, *2020*, e6765274. [CrossRef]
16. Zhao, W.; Chen, F.; Huang, H.; Li, D.; Cheng, W. A New Steel Defect Detection Algorithm Based on Deep Learning. *Comput. Intell. Neurosci.* **2021**, *2021*, e5592878. [CrossRef]
17. Jeon, Y.-J.; Choi, D.; Lee, S.J.; Yun, J.P.; Kim, S.W. Steel-surface defect detection using a switching-lighting scheme. *Appl. Opt.* **2016**, *55*, 47–57. [CrossRef] [PubMed]
18. Howard, A.G.; Zhu, M.; Chen, B.; Kalenichenko, D.; Wang, W.; Weyand, T.; Andreetto, M.; Adam, H. MobileNets: Efficient Convolutional Neural Networks for Mobile Vision Applications. *arXiv* **2017**, arXiv:1704.04861. [CrossRef]
19. Sandler, M.; Howard, A.; Zhu, M.; Zhmoginov, A.; Chen, L.-C. MobileNetV2: Inverted Residuals and Linear Bottlenecks. In Proceedings of the IEEE Conference on Computer Vision and Pattern Recognition, Salt Lake City, UT, USA, 18–23 June 2018; pp. 4510–4520. Available online: https://openaccess.thecvf.com/content_cvpr_2018/html/Sandler_MobileNetV2_Inverted_Residuals_CVPR_2018_paper.html (accessed on 23 June 2023).
20. Zhang, X.; Zhou, X.; Lin, M.; Sun, J. ShuffleNet: An Extremely Efficient Convolutional Neural Network for Mobile Devices. In Proceedings of the IEEE Conference on Computer Vision and Pattern Recognition, Salt Lake City, UT, USA, 18–23 June 2018; pp. 6848–6856. Available online: https://openaccess.thecvf.com/content_cvpr_2018/html/Zhang_ShuffleNet_An_Extremely_CVPR_2018_paper.html (accessed on 23 June 2023).
21. Iandola, F.N.; Han, S.; Moskewicz, M.W.; Ashraf, K.; Dally, W.J.; Keutzer, K. SqueezeNet: AlexNet-level accuracy with 50× fewer parameters and <0.5 MB model size. *arXiv* **2016**, arXiv:1602.07360. [CrossRef]
22. Choi, S.; Choi, J. Arithmetic Intensity Balancing Convolution for Hardware-aware Efficient Block Design. *arXiv* **2023**, arXiv:2304.04016. [CrossRef]
23. Hu, J.; Shen, L.; Sun, G. Squeeze-and-Excitation Networks. In Proceedings of the IEEE Conference on Computer Vision and Pattern Recognition, Salt Lake City, UT, USA, 18–23 June 2018; pp. 7132–7141. Available online: https://openaccess.thecvf.com/content_cvpr_2018/html/Hu_Squeeze-and-Excitation_Networks_CVPR_2018_paper.html (accessed on 23 June 2023).
24. Wan, C.; Ma, S.; Song, K. TSSTNet: A Two-Stream Swin Transformer Network for Salient Object Detection of No-Service Rail Surface Defects. *Coatings* **2022**, *12*, 1730. [CrossRef]
25. Bergstrom, A.C.; Conran, D.; Messinger, D.W. Gaussian Blur and Relative Edge Response. *arXiv* **2023**, arXiv:2301.00856. [CrossRef]
26. Lindeberg, T. Scale-Space Theory: A Basic Tool for Analyzing Structures at Different Scales. *J. Appl. Stat.* **1994**, *21*, 225–270. [CrossRef]
27. Guo, Q.; Wu, X.-J.; Kittler, J.; Feng, Z. Self-grouping convolutional neural networks. *Neural Netw.* **2020**, *132*, 491–505. [CrossRef] [PubMed]
28. Liu, W.; Anguelov, D.; Erhan, D.; Szegedy, C.; Reed, S.; Fu, C.-Y.; Berg, A.C. SSD: Single Shot MultiBox Detector. *arXiv* **2015**, arXiv:1512.02325. Available online: https://arxiv.org/abs/1512.02325v5 (accessed on 25 June 2023).
29. Lowe, D.G. Object recognition from local scale-invariant features. In Proceedings of the Seventh IEEE International Conference on Computer Vision, Corfu, Greece, 20–27 September 1999; Volume 2, pp. 1150–1157. [CrossRef]
30. Simoncelli, E.P.; Freeman, W.T. The steerable pyramid: A flexible architecture for multi-scale derivative computation. In Proceedings of the International Conference on Image Processing, Washington, DC, USA, 23–26 October 1995; Volume 3, pp. 444–447. [CrossRef]
31. El-Sennary, H.A.E.-F.; Hussien, M.E.; Ali, A.E.-M.A. Edge Detection of an Image Based on Extended Difference of Gaussian. *Am. J. Comput. Sci. Technol.* **2019**, *2*, 35–47. [CrossRef]
32. Srivastava, N.; Hinton, G.; Krizhevsky, A.; Sutskever, I.; Salakhutdinov, R. Dropout: A Simple Way to Prevent Neural Networks from Overfitting. *J. Mach. Learn. Res.* **2014**, *15*, 1929–1958.
33. Zhang, C.; Hu, H.; Fang, D.; Duan, J. The CCD sensor video acquisition system based on FPGA&MCU. In Proceedings of the 2020 IEEE 9th Joint International Information Technology and Artificial Intelligence Conference (ITAIC), Chongqing, China, 11–13 December 2020; pp. 995–999. [CrossRef]
34. Loshchilov, I.; Hutter, F. Decoupled Weight Decay Regularization. *arXiv* **2019**, arXiv:1711.05101. [CrossRef]

Disclaimer/Publisher's Note: The statements, opinions and data contained in all publications are solely those of the individual author(s) and contributor(s) and not of MDPI and/or the editor(s). MDPI and/or the editor(s) disclaim responsibility for any injury to people or property resulting from any ideas, methods, instructions or products referred to in the content.

Article

WFRE-YOLOv8s: A New Type of Defect Detector for Steel Surfaces

Yao Huang, Wenzhu Tan, Liu Li * and Lijuan Wu *

College of Physical Science and Technology, Shenyang Normal University, Shenyang 110034, China; 18360986876@163.com (Y.H.); synu163@163.com (W.T.)
* Correspondence: liliu2020@synu.edu.cn (L.L.); wulijuan@synu.edu.cn (L.W.)

Abstract: During the production of steel, in view of the manufacturing engineering, transportation, and other factors, a steel surface may produce some defects, which will endanger the service life and performance of the steel. Therefore, the detection of defects on a steel surface is one of the indispensable links in production. The traditional defect detection methods have trouble in meeting the requirements of high detection accuracy and detection efficiency. Therefore, we propose the WFRE-YOLOv8s, based on YOLOv8s, for detecting steel surface defects. Firstly, we change the loss function to WIoU to address quality imbalances between data. Secondly, we newly designed the CFN in the backbone to replace C2f to reduce the number of parameters and FLOPs of the network. Thirdly, we utilized RFN to complete a new neck RFN to reduce the computational overhead and, at the same time, to fuse different scale features well. Finally, we incorporate the EMA attention module into the backbone to enhance the extraction of valuable features and improve the detection accuracy of the model. Extensive experiments are carried out on the NEU-DET to prove the validity of the designed module and model. The mAP0.5 of our proposed model reaches 79.4%, which is 4.7% higher than that of YOLOv8s.

Keywords: surface defect detection; YOLOv8; WIoU; CFN; deep learning

1. Introduction

Steel is one of the most important industrial materials and is widely applied in the manufacture of various industrial products; therefore, quality inspection of steel is essential. During the smelting process, steel is susceptible to various defects caused by various external factors, which can affect the performance and life of the steel [1–3]. Traditional surface defect detection methods contain electromagnetic acoustic transducers, ultrasonic testing, and X-ray inspection. However, this method is inefficient, and it could result in less reliable results due to the experience of the inspector. Therefore, with the rapid advancement of machine vision, the industry is beginning to introduce machine vision technology into the detection of steel surface defects, which replaces the traditional surface defect detection method. However, conventional machine learning depends heavily on manual design algorithms in feature extraction. This could result in defect detection methods that lack versatility and robustness [4].

Recently, deep learning-based object detection algorithms are developing rapidly, and a great deal of excellent target detection models have emerged. More and more researchers try to use target detection models to detect different defect types, not only steel surface defects but also PCB solder joints [5–7], automotive paint detection [8,9], and so on. Deep learning-powered defect detection methods are separated into one-stage algorithms and two-stage algorithms. The one-stage algorithms solve the object detection as a regression problem, mainly SSD [10], YOLOv1 [11], YOLOv2 [12], YOLOv3 [13], YOLOv4 [14], YOLOv5 [15], YOLOv6 [16], YOLOv7 [17], and so on. The two-stage algorithms utilize selective search algorithms or region suggestion networks for object detection, such as

Citation: Huang, Y.; Tan, W.; Li, L.; Wu, L. WFRE-YOLOv8s: A New Type of Defect Detector for Steel Surfaces. *Coatings* **2023**, *13*, 2011. https://doi.org/10.3390/coatings13122011

Academic Editor: Yuri M. Strzhemechny

Received: 5 October 2023
Revised: 9 November 2023
Accepted: 25 November 2023
Published: 28 November 2023

Copyright: © 2023 by the authors. Licensee MDPI, Basel, Switzerland. This article is an open access article distributed under the terms and conditions of the Creative Commons Attribution (CC BY) license (https://creativecommons.org/licenses/by/4.0/).

R-CNN [9], Fast R-CNN [18], Faster R-CNN [19], R-FCN [20], and so on. They have the advantage of high accuracy and the disadvantage of being slow. In contrast, one-stage algorithms have the advantage of achieving a balance between accuracy and speed. They are easier to deploy on embedded devices.

In this study, a new steel surface defect detector, called WFRE-YOLOv8s, for detecting steel surface defects, which is based on YOLOv8s, is proposed. WFRE-YOLOv8s redesigns the backbone by utilizing the CFN module and EMA attention to reduce the number of parameters while enhancing the capability of the feature extraction. Besides that, the neck is improved by proposing a new module, RFN, to better fuse features at different scales. The main work is as follows:

1. The WIoU is employed as the loss function of WFRE-YOLOv8s. It effectively balances the gap between high-quality and low-quality data in steel surface defect datasets.
2. We have developed a CFN module that replaces the C2f module in the backbone, enhancing network detection accuracy and detection speed. Additionally, it minimizes the number of parameters and FLOPs within the entire network.
3. We have newly designed a neck, named RFN, to reduce the computational overhead. It can fuse different scale features, thus improving the accuracy of the whole detection network.
4. We have incorporated the EMA into the backbone to optimize the capacity for the extraction of valuable features for steel surface defects. This enhancement has been introduced without any additional load on the network, resulting in increased accuracy in defect detection.
5. We carry out a series of experiments primarily on NEU-DET and GC10-DET. The experimental outcomes demonstrated that our proposed methodology yields superior detection results.

2. Related Works

The defect detection methods have been comprehensively divided into conventional machine learning methods and deep learning-powered methods.

2.1. Conventional Machine Learning Methods

Machine learning has played an essential role in defect detection, and there are still many organizations that use machine learning methods to inspect their products. Franz [21] proposed using a Bayesian network classifier to detect surface defects on rough steel blocks. This method can effectively classify the defects, and the accuracy can reach 98%. Yun [22] proposed using the undecimated wavelet transform and vertical projection profile for detecting vertical line defects. Song et al. [23] proposed a new detection method incorporating saliency linear scanning morphology. This involved extracting visual saliency to eliminate background clutter and applying morphology edge processing to eliminate oil pollution edges.

Tian et al. [24] devised an enhanced ELM machine learning algorithm, incorporating a genetic algorithm, which they employed to detect surface defects on hot-rolled steel plates. Wang et al. [25] presented an improved random forest algorithm with the optimal multi-feature set fusion (OMFF-RF algorithm) for distributed defect recognition on steel surfaces. Gong et al. [26] proposed a novel multi-hypersphere support vector machine (MHSVM+) with additional information for multi-class steel surface defect classification. Chu et al. [27] developed multi-informative twin support vector machines (MTSVMs) based on binary twin support vector machines to detect steel surface defects. Zhang et al. [28] proposed a method that involves merging the Gaussian function, which is fitted to the histogram of the testing image, with the membership matrix to identify and diagnose defects. Ji et al. [29] proposed an MGH, a hybrid method utilizing machine learning and genetic algorithms, for assessing the quality of hot-rolled steel strips in production systems.

2.2. Deep Learning Approaches

The advancement of deep learning has led to the use of convolutional neural networks for target detection tasks that cannot be handled by machine learning. Object detection algorithms based on deep learning have been categorized as one-stage algorithms and two-stage algorithms. The majority of defect detection networks rely on target detection networks, while only a small portion utilize segmentation algorithms.

Bulnes [30] developed a novel defect detection technique utilizing a genetic algorithm to optimize configuration parameters. Additionally, a neural network is used for defect classification. Guan et al. [31] used VGG19 for pre-training, SVM (support vector machine), and decision trees to assess feature images' quality. Then they adjusted the parameters and structure of VGG19, thus obtaining a new VSD network for classifying steel surface defects. Xiao et al. [32] developed an image pyramid convolutional neural network (IPCNN) model based on Mask-R CNN to detect surface defects in images. Zhao et al. [33] used deformable convolution in Faster R-CNN and introduced a feature pyramid network to obtain an improved Faster R-CNN network for steel surface defect detection.

Zhao et al. [34] proposed a variant of YOLOv5L, called RDD-YOLO, to identify steel surface defects. It changed the original backbone component to Res2Net based on YOLOv5 and designed a dual feature pyramid network (DFPN) to deepen the network. Additionally, this approach utilizes a decoupling header to separate the regression and classification to improve the precision. Wang et al. [35] proposed a variant of YOLOv5s, called multi-scale-YOLOv5, to complete the detection of steel surface defects. Li et al. [36] proposed a variant of YOLOv4 for detecting defects on steel strip surfaces, which improves the precision of detection by incorporating the CBAM, where the SPP module is replaced with the RFB module. Liu et al. [37] proposed a DLF-YOLOF for defect detection on steel plate surfaces, which uses an anchorless detector to reduce the hyperparameters, utilizes a deformable convolutional network and a local spatial attention module to expand the contextual information in the feature maps, and employs a soft non-maximal suppression to improve the detection accuracy. Wang et al. [38] proposed a unique method, which is based on YOLOv7, to improve the accuracy of detecting strip steel surface defects. The ConvNeXt module has been integrated into the backbone while the attention mechanism has been incorporated into the pooling layer to enhance the ability of YOLOv7 to extract features and identify small features. Shao et al. [39] proposed a steel surface defect detection model based on a multi-scale lightweight network. This network can effectively reduce the number of parameters while achieving better model accuracy and efficiency. Inspired by YOLOv8, we propose a model named WFRE-YOLOv8s to improve detection accuracy and reduce the number of parameters and FLOPs. Compared with YOLOv8s, WFRE-YOLOv8s significantly improves prediction accuracy and identifies a wider range of defects.

3. Methods

3.1. The YOLOv8 Algorithm

3.1.1. YOLO Algorithm

The YOLO is a one-stage object detection algorithm that not only focuses on accuracy but also speed. YOLO consists of four parts: input, backbone, neck, and head. Whether it is YOLOv3, YOLOv4, or even the latest YOLOv8, their overall architecture is similar without much change. The specific detection and recognition process of YOLOv8 for the object is shown in Figure 1. The image is scaled to the appropriate size, then it is input into the CNN. The location, size, and class of the detector are obtained through backbone, neck, and head, and the loss function is utilized to calculate the gap between the predicted frame and the real frame. The gradient descent iteration is used to narrow the gap between the predicted frame and the real frame. Finally, the weight matrix and deviation at the minimum loss function in the total number of iterations are taken to get the prediction information of the object to be detected.

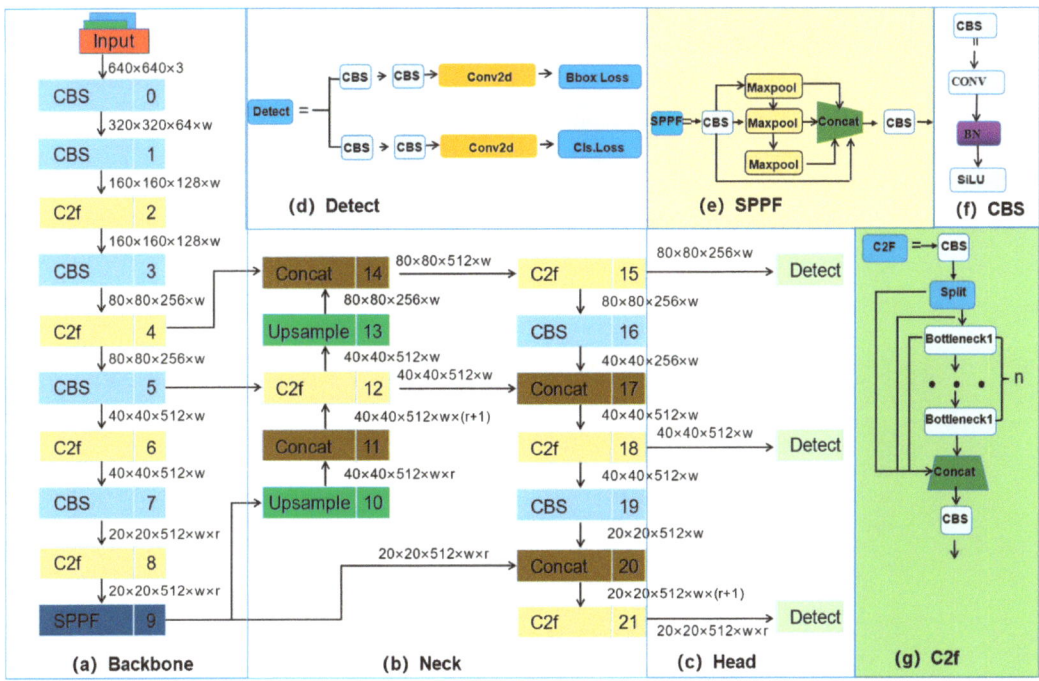

Figure 1. YOLOv8s structure. The w is the width of network and r is the ratio. (**a**–**c**) represent the structure of YOLOv8s. (**d**–**g**) represent the structure of Detect, SPPF, CBS, and C2f.

3.1.2. The Structure of YOLOv8

The YOLO algorithm has been iterated for several versions, and on 10 January 2023, Ultralytics, Inc. released YOLOv8, which is another upgrade to the many YOLO algorithms that preceded it. The YOLOv8 algorithm is similar to the YOLOv3 and YOLOv5 algorithms. It contains five versions, namely YOLOv8n, YOLOv8s, YOLOv8m, YOLOv8l, and YOLOv8x, and among these five models, YOLOv8n is the smallest and YOLOv8x is the largest. The performance differences of these models are shown in Table 1 below.

Table 1. The performance comparison of different models of YOLOv8.

Model	Size (Pixels)	mAP (0.5:0.95)	Parameters (M)	FLOPs (G)	Speed (ms)
YOLOv8n	640 × 640	37.3	3.2	8.7	0.99
YOLOv8s	640 × 640	44.9	11.2	11.2	1.20
YOLOv8m	640 × 640	50.2	25.9	25.9	1.83
YOLOv8L	640 × 640	52.9	43.7	43.7	2.39
YOLOv8x	640 × 640	53.9	68.2	68.2	3.53

From Table 1, it is evident that YOLOv8n has the fastest detection speed among these five models. Additionally, it has the lowest amount of FLOPs and number of parameters. In contrast, YOLOv8x has the slowest detection speed among these five models and a higher amount of FLOPs and number of parameters. The difference between the models is due to their sizes. The YOLOv8 algorithm mainly consists of input, backbone, neck, and head.

The structure of YOLOv8s is shown in Figure 1:

1. The backbone is utilized for feature extraction and consists of the CBS, C2f, and SPPF modules. The CBS makes a convolution operation on the input information, applies batch normalization, and activates the information stream by SiLU activation. C2f module replaces the C3 module in YOLOv5 for residual feature learning, which enriches the information stream of the feature extraction network while maintaining a lighter weight compared to C3. The SPPF module is the same as in YOLOv5, which converts arbitrary feature maps into fixed-size feature vectors.
2. The neck adopts the structure of FPN + PAN to realize the fusion between multi-scale information. Compared to YOLOv5, C3 was updated to C3.
3. The head is utilized to output the coordinates of the predicted box and the confidence of each category. Compared with YOLOv5, this part adopts a more advanced decoupled head (decoupled head). The decoupled head makes use of two independent branches to complete the task of object classification and location prediction and uses different loss functions in these two branches.

3.2. Improvement of YOLOv8s Network

In the pursuit of greater precision in detecting steel surface defects, we propose our novel model called WFRE-YOLOv8s, which is delineated in detail in Figure 2. The backbone mainly consists of CBS, SPPF, the newly proposed CFN module in this paper, and the EMA. The CFN module is a new module specially designed to replace the C2f module for steel inspection tasks. It has the advantage of using fewer parameters and less computation than the C2f module. The neck part adopts Unsample, CBS, and the RFN module in this paper. The head part is unchanged, and the WIoU is adopted as the loss function of WFRE-YOLOv8s. The RFN module is also utilized to design a new neck for the first time, which gives the model faster detection speed and higher accuracy.

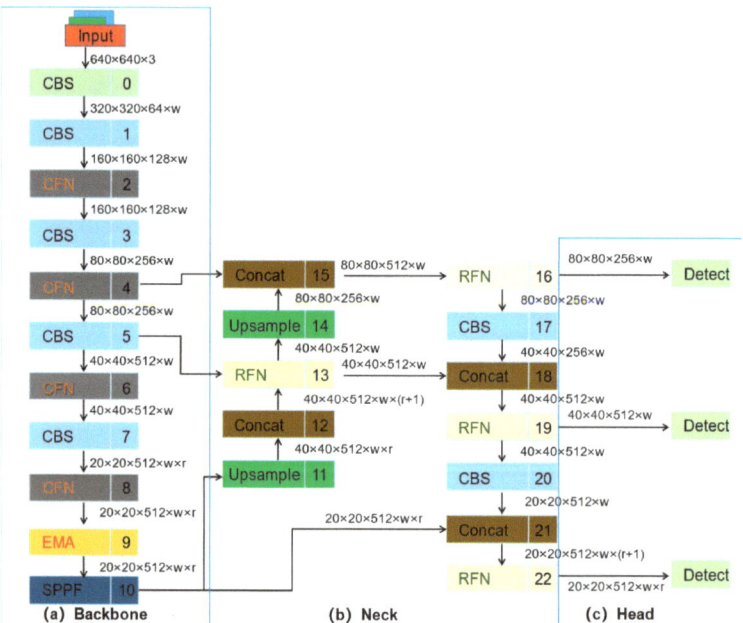

Figure 2. WFRE-YOLOv8s structure. CFN and RFN are proposed for the first time. The EMA is introduced into the backbone for the first time.

3.2.1. Improvement of the Loss Function

YOLOv8 utilizes a blend of DFL [40] and CIoU (Complete-IoU) [41] for the regression loss. However, CIoU does not take into account the balance between complex and easy samples. It also ignores the discrepancies between the bounding box and the ground truth bounding box when the penalty factor is equal to aspect ratio of the bounding box and the ground truth bounding box. As a result, CIoU boosts the computational complexity of the model. Formulas (1) and (2) illustrate the expression for the CIoU and the bounding box loss function, respectively.

$$CIoU = IoU - \frac{\rho^2(b, b^{gt})}{c^2} - av \quad (1)$$

$$L_{CIOU} = 1 - IoU + \frac{\rho^2(b, b^{gt})}{c^2} + av \quad (2)$$

In Formulas (1) and (2), IoU denotes the intersection and concatenation ratio between the bounding box and the ground truth bounding box; b, b_{gt} denotes the centroid of the ground truth bounding box and the centroid of the bounding box; a is a parameter for balancing proportionality, and v is utilized for measuring the proportionality consistency between the widths and heights of the bounding box and the ground truth bounding box.

There are frequent instances of inferior samples within the dataset of steel surface defects. This paper introduces WIoU (Wise-IoU) [42] to replace the CIoU combined with DFL to form the regression loss of the WFRE-YOLOv8s algorithm. WIoU utilizes the dynamic non-maximum suppression to assess overlap between the predicted bounding box and the ground truth bounding box. This loss function effectively improves the imbalance between the high-quality and low-quality data in dataset and the accuracy of the object detection algorithm. The calculation formula of WIoU is shown in Formulas (3) and (4):

$$L_{WIoU} = R_{WIoU} \times (1 - IoU) \quad (3)$$

$$R_{WIoU} = \exp\frac{(x - x_{gt})^2 + (y - y_{gt})^2}{(W_g^2 + H_g^2)^*} \quad (4)$$

In Formula (4), x and y represent the coordinates of the centroid of the bounding box, x_{gt} and y_{gt} represent the coordinates of the centroid of the ground truth bounding box, and w_g and h_g represent the width and height of the minimum bounding box, respectively. * represents the separation operation.

3.2.2. Improvement of the Backbone

The C2f is newly proposed in YOLOv8 to replace the C3 in YOLOv5. The C2f is designed with reference to C3 and the ELAN concept to ensure that YOLOv8 is able to acquire gradient flow information efficiently while maintaining its light weight. However, the module contains more convolutional layers, which require many convolution operations, resulting in increased computation and slowing down the model's inference speed to some extent.

Chen et al. [43] proposed a novel convolution PConv, exploiting the redundancy of the feature map to optimize costs. PConv was also utilized in the FasterNet Block and FasterNet. PConv utilizes a standard convolutional operation to extract spatial features from a subset of the input channel while keeping the residual channels. This method has the benefit of reducing computational redundancy and memory access at the same time. The structure of PConv, FasterNet Block, and FasterNet are shown in Figure 3. The FasterNet Block comprises a PConv, 1 × 1 Conv, and 1 × 1 Conv. Formula (5) explains how to calculate the FLOPs of PConv.

$$FLOPs_{PConv} = h \times w \times k \times k \times c_p \times c_p \quad (5)$$

$$FLOPs_{Conv} = h \times w \times k \times k \times c \times c \tag{6}$$

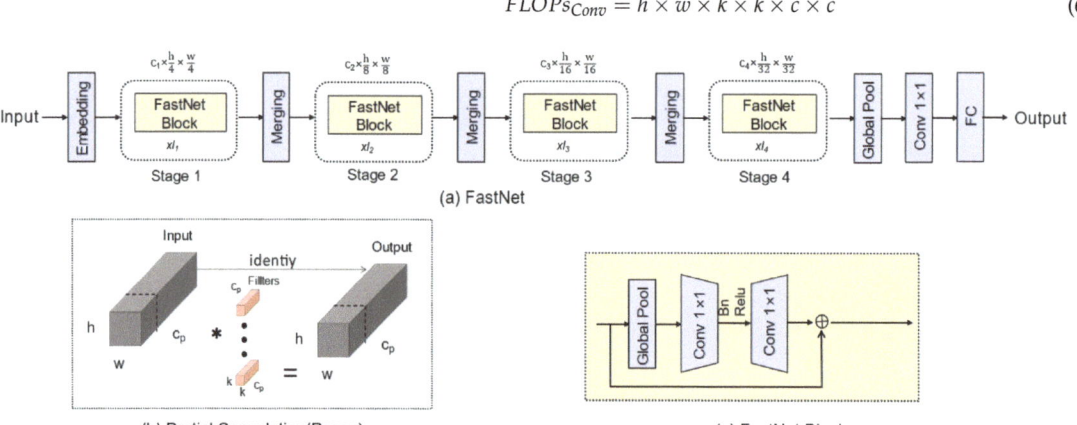

Figure 3. The structure of FastNet, FasterNet Block, and PConv. (a–c) represent the structure of FastNet, Pconv, and FastNet Block, respectively. * represent the convolution.

Formula (6) demonstrates how to calculate the FLOPs of Conv. When the ratio of c_p to the number of input feature channels c is 1/4, the FLOPs of PConv decrease to only 1/16 of those required for conventional convolution. This leads to the conclusion that PConv reduces both the FLOPs of the network and the number of parameters.

The design of a CFN module was inspired by the C3 module and FasterNet Block, illustrated in Figure 4. CFN is composed of FasterNet Block, CBS, and Concat. It differs from the traditional C3 structure in that the BottleNeck is replaced with FasterNet Block. The FasterNet Block replaces the BottleNeck in C3. Compared to the BottleNeck, this substitution improves the efficiency of feature extraction and compresses the network volume. Consequently, we replaced the C2f with the CFN to reconstruct the backbone. This approach results in a decrease in the number of parameters, FLOPs, and model size that effectively improves the model's inference speed.

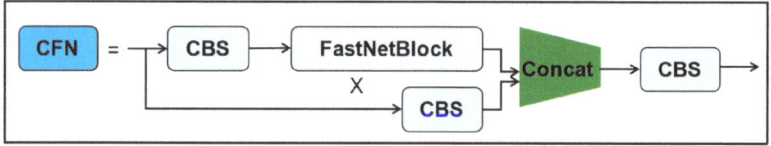

Figure 4. The structure of CFN. CFN is proposed for the first time in this paper, composed of the CBS, FastNet Block, and Concat. The core of CFN is FastNet Block; it can effectively reduce the number of FLOPs and parameters.

3.2.3. Improvement of the Neck

The neck of YOLOv8s is still a continuation of the neck structure in YOLOv5, where the FPN + PAN structure is used to complete the integration of features extracted by the backbone at varying stages to enhance the model's ability to identify features at varying scales. More and more structures about neck have been proposed to enhance the neck for full integration between multi-scale information, but also increase the computational cost simultaneously. In our study, we opted not to design novel neck modules to circumvent additional connections and fusions among feature pyramids. DAMO-YOLO [44] proposed a new EfficientRepGFPN based on GFPN, which significantly improves the accuracy of the model by utilizing various scales of feature maps for different channel dimensions in feature fusion.

The RFN is illustrated in Figure 5. Input is composed of three layers, and 1 × 1 Conv adjusts the number of channels on two parallel branches after Concat. Multiple Rep 3 × 3 Conv and 3 × 3 Conv form the efficient layer aggregation network (ElAN) [16]. RepConv is a model re-referencing technique that improves the efficiency and performance of models by merging multiple computational modules into one during the inference phase. ELAN fuses features from different layers by introducing a multi-scale feature fusion module. This can make full use of different levels of semantic information to enhance the ability to demonstrate the model's features. Due to the RFN incorporating RepConv and ELAN, the RFN can achieve much higher precision without bringing an extra computational burden.

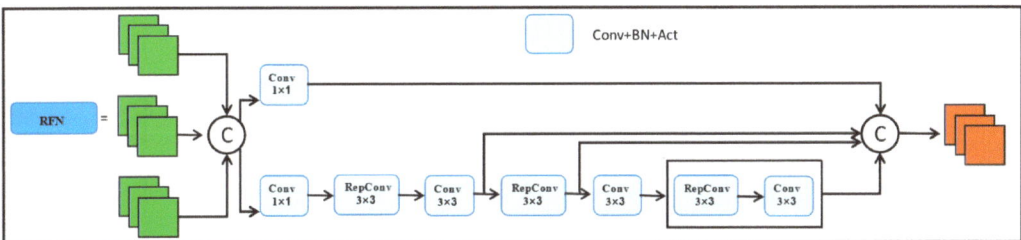

Figure 5. The structure of RFN.

Inspired by RFN, we redesigned the neck in WFRE-YOLOv8s based on the design idea of the DAMO-YOLO network and replaced the C2f with RFN in the neck part of the initial network, which brings the higher accuracy and real-time detection of the whole detection network. The improved neck is shown in Figure 6.

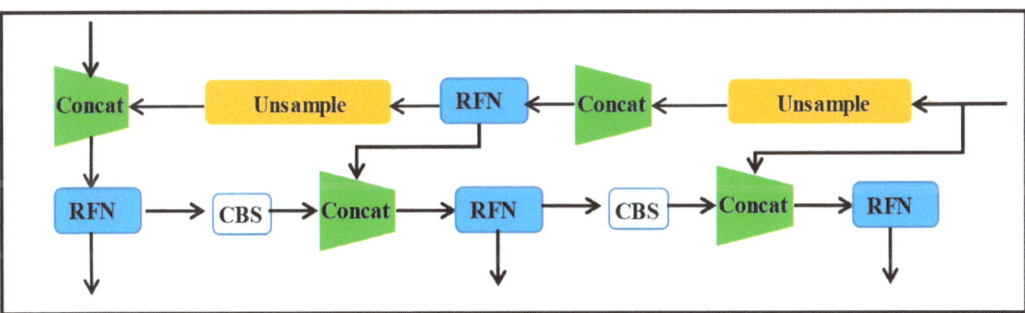

Figure 6. The improved neck.

3.2.4. Integration of EMA

Attention is widely used in computer vision, and incorporating attention into a network enables it to pay close attention to different regions of the feature map to some extent, leading to better accuracy in target identification. Attention can be mainly classified into channel attention, spatial attention, and channel spatial attention.

Due to the complexity of steel surface defects and the low pixel size of the dataset utilized, some defects are not detected or are inaccurately detected. In this paper, we will incorporate attention into the backbone to achieve higher detection accuracy. Notably, usual attention mechanisms used include CBAM [45], SE [46], ECA [47], SA [48], CA [49], and others. However, the attention models using channel dimensionality reduction to model cross-channel relationships may bring some side effects when extracting deep visual representations. Nonetheless, an efficient multi-scale attention (EMA) [50] module based on coordinate attention (CA) is proposed to better address this issue. It encodes global information to recalibrate the channel weights in each parallel branch while further aggregating the output features of the two parallel branches through cross-dimensional interaction to capture the pixel-level pairwise relationship, achieving the goal of reducing computational overhead while preserving the information of each channel. CA attention first divides the input information according to the two directions of width and height, thus obtaining the feature information of width and height. The global average pooling formulas for both are shown in Formulas (7) and (8).

$$z_c^h(h) = \frac{1}{W} \sum_{0 \leq i \leq W} x_c(h, i) \tag{7}$$

$$z_c^w(w) = \frac{1}{W} \sum_{0 \leq i \leq H} x_c(j, w) \tag{8}$$

Next, the feature maps of the global perceptual field in both the width and height directions are spliced, and feature transformations are performed using 1×1 convolution, batch normalization algorithm, and nonlinear activation. Immediately after that, the feature transformation is achieved by 1×1 convolution and Sigmoid activation function so that its dimension is the same as the input X vector, and then the attentional weights g_h and g_w are computed for the achieved feature maps in the width direction and height direction. Finally, the output g_h and g_w are combined into a weight matrix by weighted multiplication computation on the original feature maps, and the result is shown in Formula (9). CA is shown in Figure 7a.

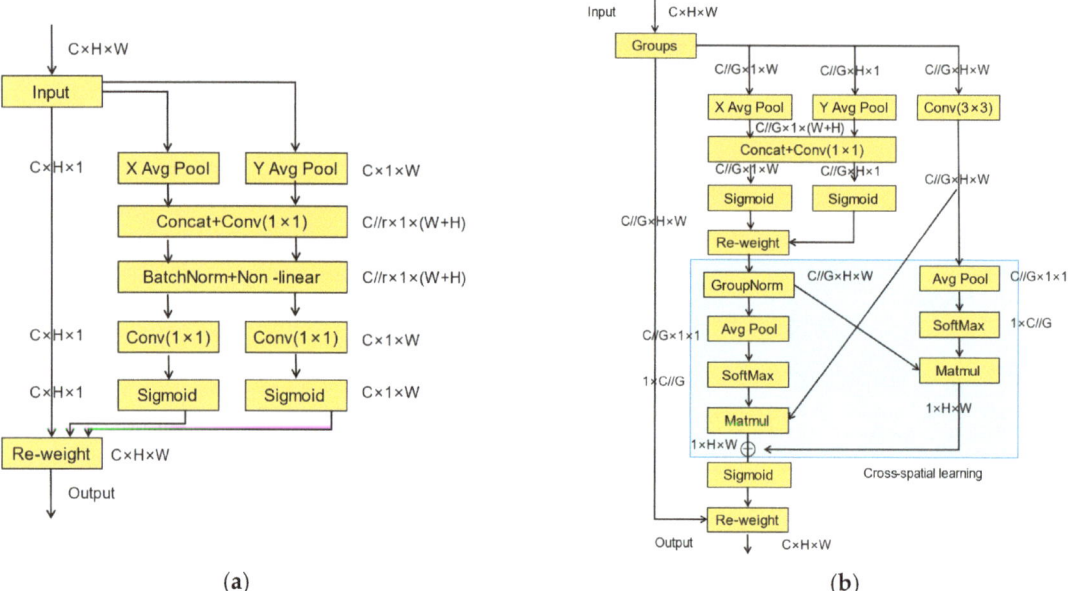

Figure 7. CA and EMA. This figure illustrates the difference in structure between CA and EMA. The structure of CA and EMA is shown in (**a**,**b**).

EMA borrows the idea of CA and designs three parallel routes to extract the attention weight descriptors of grouped feature maps. The two routes on the left, similar to CA, are named 1×1 branches and the rightmost route is named 3×3 branch.

$$y_c(i, j) = x_c(i, j) \times g_c^h(i) \times g_c^w(j) \tag{9}$$

In the 1 × 1 branch, similar to CA attention, the X and Y global average pooling module is used to extract feature information in the width and height directions, the feature information is spliced, and 1 × 1 convolution is used to prevent dimensionality reduction. The 3 × 3 branch utilizes the 3 × 3 convolution to capture local cross-channel interactions to expand the feature space.

In the cross-space learning module, the global spatial information in the output of the 1 × 1 branch is first encoded using global average pooling, after which a Softmax function is fitted to the linear transform to ensure efficient computation. Finally, the output of the parallel processing is multiplied by the matrix dot product operation to obtain the first spatial attention map.

Then, the second spatial attention map retaining the exact spatial location information is obtained by employing global average pooling and fitting a linear transformation with the Softmax function at the 3 × 3 branch. Finally, the output feature maps for each group are calculated by summing the two spatial attention weight values that were generated using the sigmoid function. The global average pooling operates as shown in Formula (10).

$$z_c(h) = \frac{1}{H \times W} \sum_{j}^{H} \sum_{i}^{W} x_c(i,j) \qquad (10)$$

As shown in Figure 8, the EMA attention is incorporated prior to the spatial pyramid pooling module of the YOLOv8s backbone network. This integration boosts the accuracy of WFRE-YOLOv8s for identifying surface flaws in steel. Additionally, this enhancement is accomplished without placing additional computational burden on the network infrastructure.

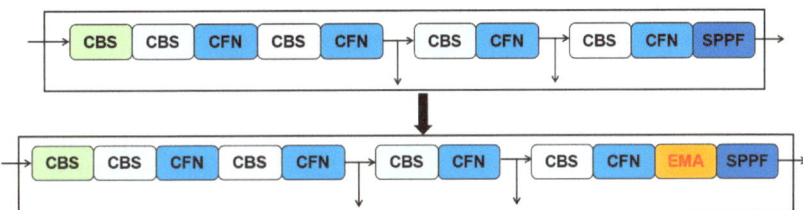

Figure 8. EMA module added to the backbone network. This figure shows where EMA is located in the backbone network.

4. Experiments

4.1. Experimental Introduction

4.1.1. Experimental Setup

The running environment of the experiment is as follows: the operating system is Windows 10 Professional, the CPU is Intel I5-12490F, the GPU is NVIDIA GeForce RTX3060 12 G, and the RAM is 16 G. Some specific functions may be missing between versions causing the environment to crash. The version of Python and the version of Torch, CUDA, and CUDNN must match, or else the model will not start running. Therefore, the Python environment is based on Anaconda's Python 3.8, the Pytorch version is 1.9, the CUDA version is 11.5, and the CUDNN version is 8005.

The specific training parameters are set as follows: the image size is 640 × 480, the initial learning rate is 0.01, the number of iterations is 200, the batch size is 16, Num_Workers is 2, and the mosaic enhancement is turned off after 190 epoch.

4.1.2. Evaluation Indicators

In this study, mAP (mean average precision), recall, precision, parameters, and FLOPs were used to evaluate the performance of the improved algorithm. The formula for calculating the mean values of recall, precision, and mAP are shown:

$$Recall = \frac{TP}{TP + FN} \tag{11}$$

$$Precision = \frac{TP}{TP + FP} \tag{12}$$

$$AP = \int_0^1 P(R)dR \tag{13}$$

$$mAP = \frac{1}{N}\sum_{i=1}^{N} AP_i \tag{14}$$

where TP represents the number of road objects predicted correctly, FP represents the number of road objects predicted incorrectly, and FN represents the number of road objects missed. P(R) represents the value of precision under the point recall.

4.1.3. Dataset

Due to the fact that the images in the NEU-DET dataset are derived from the real steelmaking process, its defects are closer to the actual situation; this is the reason why we use the NEU-DET to validate the effectiveness of WFRE-YOLOv8s. GC10-DET contains a wider variety of metal surface defects, it can help validate the model's versatility and robustness across a broader range of defect categories. We applied mosaic image enhancement to the NEU-DET and GC10-DET. NEU-DET [51] was proposed by the team of Kechen Song at Northeastern University in 2022. NEU-DET includes six types of steel surface defects, namely, patches (pa), silvering (cr), inclusions (in), roll marks (rs), scratches (sc), and pitting surfaces (ps). The number of images per defect category is 300 and the resolution of each image is 200 × 200. The ratio of the training set to the validation set of NEU-DET is set as 9:1, which means there are 1620 images for training and 180 images for validation. GC10-DET [52] was proposed by the team of Xiaoming Lv in 2020. GC10-DET consists of ten types of metallic surface defects, namely, water spot (ws), punching (pu), silk spot (ss), crescent gap (cg), oil spot (os), waist folding (wf), inclusion (in), rolled pit (rp), crease (cr), and weld line (wl). The GC10-DET has 2294 images, and the resolution of each image is 2048 × 1000. The ratio of the training set to the validation set of GC10-DET is set as 9:1, which means there are 2064 images for training and 230 images for validation.

4.1.4. Experimental Datum Processing

After completing training for each model, a result folder is generated, and a new folder for organizing experimental data is created on the computer. Subclass folders are then created based on different models. To ensure data authenticity, we conducted 20 training sessions for each model and uniformly stored the results in the subclass folder. The final experimental result is determined as the closest result to the average.

4.2. Comparisons with Related Methods

4.2.1. Comparisons with Prevailing Methods on NEU-DET

To evaluate the effectiveness of WFRE-YOLOv8s, we compare our method with several mainstream methods. The YOLO algorithms are now commonly used in various fields. At present, most researchers have also adopted the YOLO algorithm to design steel defect detectors. Furthermore, WFRE-YOLOv8s is based on YOLOv8, so in order to ensure the objectivity of the experiments, we chose these algorithms as the comparative objects. These methods include YOLOv3s, YOLOv4, YOLOv5s, YOLOv7, YOLOv8s, YOLOv8m, and YOLOv8L.

It is evident from the data presented in Figure 9b and Table 2 that the model introduced in this study outperforms others in terms of mAP0.5 and mAP0.95. Compared to YOLOv8s, the mAP0.5 and mAP0.95 have increased by 4.7% and 2.3%, respectively, while the number of parameters and FLOPs have only increased by 20% and 13%. Compared to YOLOv8L, which has the highest accuracy among other models, WFRE-YOLOv8s is 1.2% and 1.4% higher on mAP0.5 and mAP0.95, respectively, and the number of parameters and the FLOPs in WFRE-YOLOv8s are 68.5% and 80.3% lower than in YOLOv8L. Compared to the well-known YOLOv3s and YOLOv5, our proposed WFRE-YOLOv8s is higher than these two models by 27.5% and 20.7% and 9.9% and 7.7% on mAP0.5 and mAP0.95, respectively. Furthermore, WFRE-YOLOv8s outperforms YOLOv4 and YOLOv7 in terms of mAP0.5 and mAP0.95 while utilizing significantly fewer parameters and FLOPs. Additionally, the mAP0.5 of WFRE-YOLOv8s is 10% higher than that of YOLOv4 and 5.7% higher than that of YOLOv7. Therefore, the WFRE-YOLOv8s proposed in this paper can achieve good accuracy and maintain good detection results with the average number of parameters and computational effort.

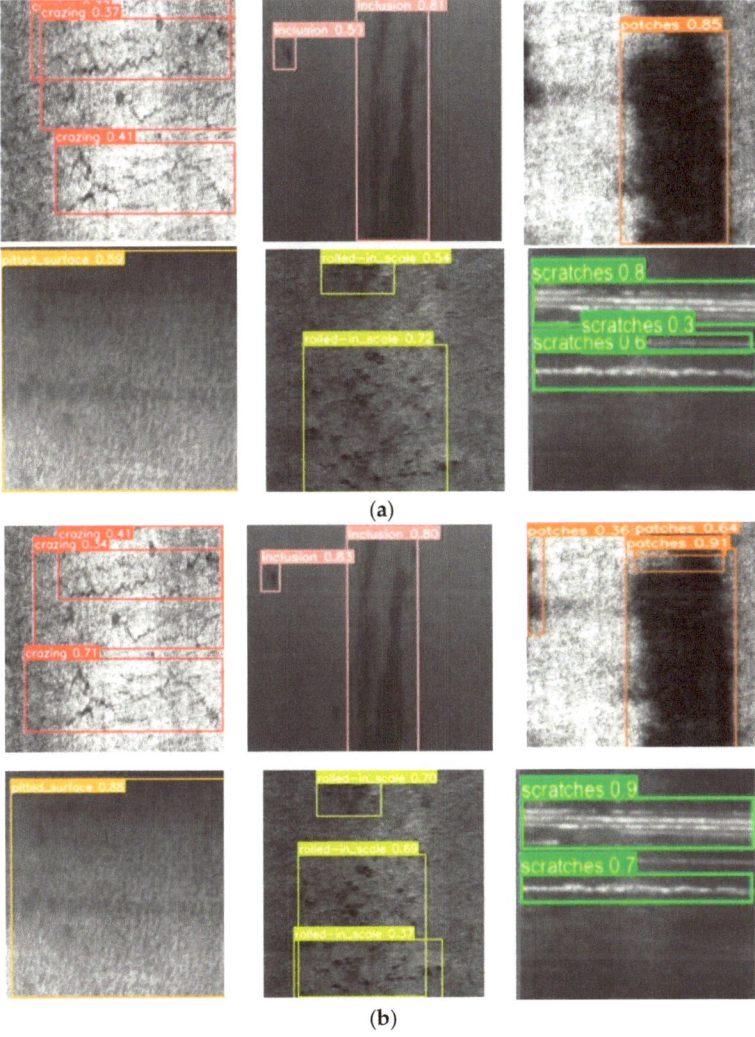

Figure 9. Comparison of detection results on NEU-DET. (**a**) YOLOv8s recognition effect. (**b**) Improved recognition of WFRE-YOLOv8s. (**a**) shows the detection results of YOLOv8s on NEU-DET. (**b**) shows the detection results of WFRE-YOLOv8s on NEU-DET.

Table 2. Comparison of different network performances on NEU-DET. This table illustrates the experimental results of the different methods on NEU-DET. (The indicator of focus is mAP (0.5)).

Model	mAP (0.5)	mAP (0.5:0.95)	Recall	Precision	Parameters	FLOPs (G)
YOLOv3s	51.9%	21.8%	45.4%	49.5%	9322387	23.4
YOLOv4	69.4%	33.8%	79.3%	53.1%	63964611	142
YOLOV5s	69.5%	34.8%	72.7%	60.8%	7026307	15.8
YOLOv7	73.7%	34.4%	68.6%	66.3%	37223526	105.2
YOLOv8s (baseline)	74.7%	39.4%	69.0%	69.1%	11127906	28.4
YOLOv8m	76.7%	41.1%	69.2%	74.5%	25843234	78.7
YOLOv8L	78.2%	41.1%	71.0%	75.8%	43611234	164.8
WFRE-YOLOv8s (ours)	79.4%	42.5%	75.9%	73.6%	13775472	32.6

4.2.2. Comparisons with Prevailing Methods on GC10-DET

For further exploration of the effectiveness of WFRE-YOLOv8s, we conducted some experiments on GC10-DET. The results of detection are shown in Table 3. We can still observe a 3.8% increase of mAP0.5 in favor of WFRE-YOLOv8s over YOLOv8s, indicating that WFRE continues to outperform YOLOv8s. Meanwhile, the number of parameters and FLOPs have only increased by 23.8% and 14.3%. Compared to YOLOv3s and YOLOv5s, WFRE-YOLOv8s is still superior to both models in terms of mAP0.5 and mAP0.95. The mAP0.5 of WFRE-YOLOv8s is 8.9%, 5.2%, 2.4%, and 0.8% higher than YOLOv4, YOLOv7, YOLOv8m, and YOLOv8L, respectively. From the above data, WFRE-YOLOv8s is not only effective for detecting the six types of defects in NEU-DET but also has good effectiveness for detecting the ten types of defects in GC10-DET.

Table 3. Comparison of different network performances on GC10-DET. This table illustrates the experimental results of the different methods on GC10-DET. (The indicator of focus is mAP (0.5)).

Model	mAP (0.5)	mAP (0.5:0.95)	Recall	Precision	Parameters	FLOPs (G)
YOLOv3s	55.0%	27.3%	58.0%	51.4%%	9333175	23.4
YOLOv4	60.5%	30.4%	68.0%	46.1%	8081831	20.6
YOLOv5s	63.3%	33.0%	54.7%	69.2%	7037095	15.8
YOLOv7	64.2%	32.1%	57.8%	69.3%	37245102	105.3
YOLOv8s (baseline)	65.6%	34.9%	61.6%	29.9%	11129454	28.5
YOLOv8m	67.0%	35.0%	55.5%	81.0%	25845550	78.7
YOLOv8L	68.6%	36.5%	73.0%	60.7%	43637550	165.48
WFRE-YOLOv8s (ours)	69.4%	35.7%	62.6%	64.8%	13775472	32.6

4.2.3. Experimental Comparison on Different Datasets

For further exploration of the versatility and robustness of WFRE on different classes of defects, we also conducted experiments on the Lv-DET [53] and PKU-Market-PCB datasets [54], and the specific experimental results as well as the performance are shown in Table 4.

Table 4. Comparison of experimental results for different datasets. This table shows a comparison of the experimental results of YOLOv8s and WFRE-YOLOv8s on different defective datasets.

Model	Model	mAP (0.5)	mAP (0.5:0.95)
NEU-DET	YOLOv8s	74.7%	39.4%
	WFRE-YOLOv8s	79.4%	42.5%
GC10-DET	YOLOv8s	65.6%	34.9%
	WFRE-YOLOv8s	69.4%	35.7%
Lv-DET	YOLOv8s	56.4%	34.5%
	WFRE-YOLOv8s	59.2%	35.4%
PCB	YOLOv8s	82.0%	42.7%
	WFRE-YOLOv8s	85.3%	44.0%

From Table 4, we can find that WFRE-YOLOv8s outperforms YOLOv8s on different datasets. On NEU-DET and GC10-DET, the mAP0.5 of WFRE-YOLOv8s is 4.7% and 3.8% higher than YOLOv8s. On the other hand, our models have achieved excellent results in the detection of non-steel defects. In aluminum defect detection, the mAP0.5 of WFRE-YOLOv8 is 2.8% higher YOLOv8s. In pcb defect detection, the mAP0.5 of WFRE-YOLOv8s is 3.3% higher than YOLOv8s. From the above comparative experimental results, our proposed model has good versatility and robustness.

4.3. Ablation Experiments

To verify whether our proposed method is stable and effective, we conducted ablation experiments on the NEU-DET, and the specific experimental comparison results are shown in Table 5 and Figure 9. The YOLOv8s model is the baseline model. Firstly, in order to validate the effectiveness of the WIoU loss function in the model proposed in this paper, the WIoU loss function is used to replace the CIoU loss function, and the model is named W-YOLOv8s. Secondly, the CFN module proposed in this paper is utilized to replace the C2f module in the backbone, and the model is named WF-YOLOv8s. Thirdly, the RFN module is introduced into the neck, and the module is utilized to replace the C2f module, and the model is named WFR-YOLOv8s. Finally, an EMA attentional mechanism is incorporated into the WFR-YOLOv8s. The addition of EMA improves the extraction of valuable features and overall detection accuracy of the whole model. The model resulting from this adjustment is dubbed WFRE-YOLOv8s.

Table 5. Ablation experiments. This table illustrates the experimental results of the different stages of the improved methodology. (The indicator of focus is mAP (0.5)).

Model	mAP (0.5)	mAP (0.5:0.95)	Recall	Precision	Parameters	FLOPs (G)
YOLOv8s	74.7%	39.4%	69.0%	69.1%	11127906	28.4
W-YOLOv8s	75.6%	40.6%	70.7%	69.8%	11127906	28.4
WF-YOLOV8s	76.6%	41.2%	71.4%	71.1%	9434466	23.5
WFR-YOLOv8s	78.1%	41.8%	72.8%	73.2%	13644386	32.2
WFRE-YOLOv8s	79.4%	42.5%	75.9%	73.6%	13775472	32.6

4.3.1. The Performance of WIoU

As shown in Table 5, after we improve the original loss function CIoU to WIoU, the mAP0.5 of the model reaches 75.6%, which is an improvement of 0.9% compared to the original model. The number of parameters for the network and the amount of computation do not change, which shows that the WIoU loss function effectively solves the imbalance between high-quality and low-quality data in the steel surface defects dataset.

4.3.2. The Performance of RFN

As can be seen from Table 5, after replacing the original C2f module with the proposed CFN module in W-YOLOv8s, the mAP0.5 is increased from 75.6% to 76.6%. The number of parameters and FLOps are decreased by 15% and 18%, respectively. This dramatically proves that the CFN module proposed in this paper has a strong compression effect on the number of parameters and FLOPs of the model. At the same time, it can also improve the detection accuracy of the model.

4.3.3. The Performance of CFN

As can be seen in Table 5, after replacing the C2f in the neck with our proposed RFN module in WF-YOLOv8s, the mAP0.5 improves by 1.5% to 78.1%, which indicates that the RFN structure can effectively enhance the detection accuracy of the network.

4.3.4. The Performance of EMA Attention

From Table 5, it can be seen that WFRE-YOLOv8s, compared to WFR-YOLOv8s, incorporates the EMA attention module in the network's backbone to enhance the network's ability to extract features of defects in low-pixel images to improve the model's detection accuracy without increasing the network's computational burden too much. The mAP0.5 increased by 1.3% to reach 79.4%, and the number of parameters and computation of the model only increased by 0.9% and 1.2%.

4.4. Comprehensive Performance of the Proposed Model

The results of detecting defects in NEU-DET using YOLOv8s and WFRE-YOLOv8s are illustrated in Figure 9. The results include predicted boxes, defect classes, and confidence scores. The results of detecting each type of defect using YOLOv8s and YOLOv7 are shown in Table 6.

Table 6. Comparison of detection results of improved algorithms.

Model	mAP (0.5)	Crazing	Inclusion	Patches	Pitted Surface	Rolled-in Scale	Scratches
YOLOv8s (baseline)	74.7%	43.6%	82.2%	94.0%	78.1%	66.8%	83.3%
WFRE-YOLOv8s (ours)	79.4%	60.0%	81.4%	93.8%	82.5%	73.8%	84.8%

From Figure 9, it is evident that our WFRE-YOLOv8s can detect crazing, patches, and rolled-scale defect targets more accurately than YOLOv8s, which missed one defect target. The target missed by YOLOv8s is well-detected, and our detection accuracy is also improved based on YOLOv8s. On the two defective targets of inclusion and pitted surface, the detection accuracy of WFRE-YOLOv8s is improved compared to YOLOv8s, and the accuracy of the same target is improved by 30% and 20%. On the defective target of scratches, the original YOLOv8s had the problem of misidentification in the detection result. However, WFRE-YOLOv8s solved this problem by accurately identifying the defects of the two scratches on this image. Figure 9 and Table 6 show that the WFRE-YOLOv8s proposed in this paper is more advanced and accurate compared to the YOLOv8s, which have better results.

4.5. Discussion

4.5.1. Findings

In our study, we found that the design of the backbone network is crucial, and the backbone network of WFRE-YOLOv8s can be reconfigured by CFN and EMA to be much more efficient than the detection of YOLOv8. The design of the neck cannot only focus on the design of the feature pyramid structure but also on the feature fusion effect of the model.

From Tables 2–5, and Figure 9, we can find that WFRE-YOLOv8s is not only better than the original model in terms of detection metrics, but also in terms of actual detection results. The newly added EMA attention better focuses on the features that would be missed by the original model, improving the model's overall detection results.

According to the experimental results, in our opinion, as much as possible, it is important to design a unique structure to focus on those features that are easily overlooked. It is also necessary to try to keep the detection network as efficient as possible, rather than trying to increase the accuracy of the network by designing more parameter-heavy structures, only to result in increased redundancy in the network. In steel production, crazing and rolled-in scale are two types of defects that are easily overlooked, so these two types of features are needed in the design of the network for the design of a targeted feature extraction structure. In terms of datasets, the current size of datasets is still far from enough; we need to increase the expansion of datasets, enrich the samples of various types of defects, and improve the quality of datasets.

4.5.2. Limitations and Future Works

Compared to existing methods, WFRE-YOLOv8s is a highly competitive steel defect detector that has performed well on NEU-DET. However, the limitations of WFRE-YOLOv8s are still apparent in Figure 9 and Table 6, as evidenced by the results of detecting defects. According to the results of detecting the crazing and rolled-in scale, we can find that its accuracy is still not as good as the other categories. This suggests that WFRE-YOLOv8s is not yet adequate in identifying these two defects and has substantial scope for improvement. It appears that the lower resolution of the dataset images and the indistinctive characteristics of these two defects may be responsible for the issue at hand. Regarding data preprocessing, it may be beneficial to employ certain image preprocessing techniques and increase the number of datasets to enhance the model's ability to detect crazing and other categories of defects.

Additionally, Table 3 reveals that while the computational effort and FLOPs of WFRE-YOLOv8s are lower compared to most YOLO models, the WFRE-YOLOv8s model still requires compression due to its use in industrial production. It is clear from Table 3 that WFRE-YOLOv8s currently has a significant number of parameters. Despite being efficient on devices with higher arithmetic capabilities, it poses a challenge when running on edge terminal devices. Given that industrial production necessitates a large number of steel defect detection sensors, increasing the deployment of high-performance computing equipment would lead to higher expenses. Thus, it is necessary to compress the number of parameters in WFRE-YOLOv8s to enhance algorithmic efficiency and lower the production costs of businesses. WFRE-YOLOv8s could benefit from being optimized further through the implementation of a lightweight convolutional backbone network, pruning, and distillation.

5. Conclusions

In response to the steel used in the production process, manual quality inspection is inefficient, the traditional machine learning quality inspection method generalization and robustness is poor, and so on. In this paper, we propose a novel one-stage detector named WFRE-YOLOv8s for steel surface defect detection, which is based on YOLOv8s with improvements in the backbone, neck, loss function, and integration of the current better EMA attention module. To solve the problem of imbalance between high-quality and low-quality data in the steel dataset, we introduce the WIoU loss function to replace the CIoU loss function. In order to reduce the amount of computation and the number of parameters in the model and keep the accuracy from being degraded, we adopt the CFN module as the main component of the backbone of WFRE-YOLOv8s. In the neck, we adopt the RFN module to reduce the computational overhead while fusing different scale features well, resulting in improved detection accuracy and real-time detection speed of the network. In addition, we also incorporate the EMA attention module in the backbone part of the play network, which can enhance the extraction of adequate feature information and thus enhance the detection accuracy of the network and solve the problem of some defects being missed and wrongly detected in the solution process. In order to validate our proposed WFRE-YOLOv8s, we conducted a series of experiments on NEU-DET and not only ablation experiments but also a comparison with other SOTA target detection models. The ablation experiments proved the effectiveness of our proposed improved module, and the comparison experiments with other SOTA models proved the effectiveness of our proposed model. In comparison with other methods, WFRE-YOLOv8s achieved better performance than other models, with higher scores than others in mAP0.5 and mAP0.95.

For WFRE-YOLOv8s, with crazing and rolled-in scale, there is still the problem that detection accuracy is lower than other categories. Additionally, there is still room for compression in terms of the number of model parameters and FLOPs. Moving forward, we will focus on enhancing the ability of the model to detect defects in all categories and designing a more lightweight model.

Author Contributions: Conceptualization, Y.H. and L.W.; methodology, Y.H. and L.W.; software, Y.H., W.T. and L.W.; validation, Y.H., L.L. and L.W.; formal analysis, Y.H., L.L. and L.W.; investigation, Y.H. and L.W.; resources, Y.H., L.L. and L.W.; data curation, Y.H.; writing—original draft preparation, Y.H.; writing—review and editing, Y.H., L.L. and L.W.; visualization, Y.H.; supervision, L.L. and L.W. All authors have read and agreed to the published version of the manuscript.

Funding: This research was funded by Scientific Research Program of Liaoning Provincial Department of Education (LFW202003).

Institutional Review Board Statement: Not applicable.

Informed Consent Statement: Not applicable.

Data Availability Statement: Data are contained within the article.

Conflicts of Interest: The authors declare no conflict of interest.

References

1. Di, H.; Ke, X.; Peng, Z.; Dongdong, Z. Surface defect classification of steels with a new semi-supervised learning method. *Opt. Lasers Eng.* **2019**, *117*, 40–48. [CrossRef]
2. Lee, S.Y.; Tama, B.A.; Moon, S.J.; Lee, S. Steel surface defect diagnostics using deep convolutional neural network and class activation map. *Appl. Sci.* **2019**, *9*, 5449. [CrossRef]
3. Xu, Y.; Li, D.; Xie, Q.; Wu, Q.; Wang, J. Automatic defect detection and segmentation of tunnel surface using modified Mask R-CNN. *Measurement* **2021**, *178*, 109316. [CrossRef]
4. Zhou, A.; Zheng, H.; Li, M.; Shao, W. Defect Inspection Algorithm of Metal Surface Based on Machine Vision. In Proceedings of the 2020 12th International Conference on Measuring Technology and Mechatronics Automation (ICMTMA), Phuket, Thailand, 28–29 February 2020; pp. 45–49.
5. Tang, J.; Liu, S.; Zhao, D.; Tang, L.; Zou, W.; Zheng, B. PCB-YOLO: An Improved Detection Algorithm of PCB Surface Defects Based on YOLOv5. *Sustainability* **2023**, *15*, 5963. [CrossRef]
6. Liao, X.; Lv, S.; Li, D.; Luo, Y.; Zhu, Z.; Jiang, C. Yolov4-mn3 for pcb surface defect detection. *Appl. Sci.* **2021**, *11*, 11701. [CrossRef]
7. Liu, Z.; Qu, B. Machine vision based online detection of PCB defect. *Microprocess. Microsyst.* **2021**, *82*, 103807. [CrossRef]
8. Zhang, J.; Xu, J.; Zhu, L.; Zhang, K.; Liu, T.; Wang, D.; Wang, X. An improved MobileNet-SSD algorithm for automatic defect detection on vehicle body paint. *Multimed. Tools Appl.* **2020**, *79*, 23367–23385. [CrossRef]
9. Kieselbach, K.K.; Nöthen, M.; Heuer, H. Development of a visual inspection system and the corresponding algorithm for the detection and subsequent classification of paint defects on car bodies in the automotive industry. *J. Coat. Technol. Res.* **2019**, *16*, 1033–1042. [CrossRef]
10. Liu, W.; Anguelov, D.; Erhan, D.; Szegedy, C.; Reed, S.; Fu, C.Y.; Berg, A.C. Ssd: Single shot multibox detector. In Proceedings of the European Conference on Computer Vision, Amsterdam, The Netherlands, 11–14 October 2016; pp. 21–37.
11. Redmon, J.; Divvala, S.; Girshick, R.; Farhadi, A. You only look once: Unified, real-time object detection. In Proceedings of the IEEE Conference on Computer Vision and Pattern Recognition, Las Vegas, NV, USA, 27–30 June 2016; pp. 779–788.
12. Redmon, J.; Farhadi, A. Yolov3: An incremental improvement. *arXiv* **2018**, arXiv:1804.02767.
13. Bochkovskiy, A.; Wang, C.Y.; Liao, H.Y.M. Yolov4: Optimal speed and accuracy of object detection. *arXiv* **2020**, arXiv:2004.10934.
14. Jocher, G. Stoken Yolov5. Available online: https://github.com/ultralytics/yolov5/releases/tag/v6.0 (accessed on 26 October 2022).
15. Li, C.; Li, L.; Jiang, H.; Weng, K.; Geng, Y.; Li, L.; Ke, Z.; Li, Q.; Cheng, M.; Nie, W.; et al. YOLOv6: A single-stage object detection framework for industrial applications. *arXiv* **2022**, arXiv:2209.02976.
16. Wang, C.Y.; Bochkovskiy, A.; Liao, H.Y.M. YOLOv7: Trainable bag-of-freebies sets new state-of-the-art for real-time object detectors. In Proceedings of the IEEE/CVF Conference on Computer Vision and Pattern Recognition, Vancouver, BC, Canada, 18–22 June 2023; pp. 7464–7475.
17. Girshick, R.; Donahue, J.; Darrell, T.; Malik, J. Rich Feature Hierarchies for Accurate Object Detection and Semantic Segmentation. In Proceedings of the IEEE Conference on Computer Vision and Pattern Recognition, New York, NY, USA, 23–28 June 2014; pp. 580–587.
18. Girshick, R. Fast r-cnn. In Proceedings of the IEEE International Conference on Computer Vision, Santiago, Chile, 7–13 December 2015; pp. 1440–1448.
19. Ren, S.; He, K.; Girshick, R.; Sun, J. Faster r-cnn: Towards Real-time Object Detection with Region Proposal Networks. *Adv. Neural Inf. Process. Syst.* **2015**, *28*, 1–9. [CrossRef] [PubMed]
20. Dai, J.; Li, Y.; He, K.; Sun, J. R-FCN: Object Detection Via Region-Based Fully Convolutional Networks. *Adv. Neural Inf. Process. Syst.* **2016**, *29*, 379–387.
21. Pernkopf, F. Detection of surface defects on raw steel blocks using Bayesian network classifiers. *Pattern Anal. Appl.* **2004**, *7*, 333–342. [CrossRef]
22. Yun, J.P.; Choi, S.; Kim, S.W. Vision-based defect detection of scale-covered steel billet surfaces. *Opt. Eng.* **2009**, *48*, 037205. [CrossRef]

23. Song, K.; Hu, S.; Yan, Y.; Li, J. Surface Defect Detection Method Using Saliency Linear Scanning Morphology for Silicon Steel Strip under Oil Pollution Interference. *ISIJ Int.* **2014**, *54*, 2598–2607. [CrossRef]
24. Tian, S.; Xu, K. An algorithm for surface defect identification of steel plates based on genetic algorithm and extreme learning machine. *Metals* **2017**, *7*, 311. [CrossRef]
25. Wang, Y.; Xia, H.; Yuan, X.; Li, L.; Sun, B. Distributed defect recognition on steel surfaces using an improved random forest algorithm with optimal multi-feature-set fusion. *Multimed. Tools Appl.* **2018**, *77*, 16741–16770. [CrossRef]
26. Gong, R.; Wu, C.; Chu, M. Steel surface defect classification using multiple hyper-spheres support vector machine with additional information. *Chemom. Intell. Lab. Syst.* **2018**, *172*, 109–117. [CrossRef]
27. Chu, M.; Liu, X.; Gong, R.; Liu, L. Multi-class classification method using twin support vector machines with multi-information for steel surface defects. *Chemom. Intell. Lab. Syst.* **2018**, *176*, 108–118. [CrossRef]
28. Zhang, J.; Wang, H.; Tian, Y.; Liu, K. An accurate fuzzy measure-based detection method for various types of defects on strip-steel surfaces. *Comput. Ind.* **2020**, *122*, 103231. [CrossRef]
29. Ji, Y.; Liu, S.; Zhou, M.; Zhao, Z.; Guo, X.; Qi, L. A machine learning and genetic algorithm-based method for predicting width deviation of hot-rolled strip in steel production systems. *Inf. Sci.* **2022**, *589*, 360–375. [CrossRef]
30. Bulnes, F.G.; Garcia, D.F.; Javier De la Calle, F.; Usamentiaga, R.; Molleda, J. A non-invasive technique for online defect detection on steel strip surfaces. *J. Nondestruct. Eval.* **2016**, *35*, 1–18. [CrossRef]
31. Guan, S.; Lei, M.; Lu, H. A steel surface defect recognition algorithm based on improved deep learning network model using feature visualization and quality evaluation. *IEEE Access* **2020**, *8*, 49885–49895. [CrossRef]
32. Xiao, L.; Wu, B.; Hu, Y. Surface defect detection using image pyramid. *IEEE Sens. J.* **2020**, *20*, 7181–7188. [CrossRef]
33. Zhao, W.; Chen, F.; Huang, H.; Li, D.; Cheng, W. A new steel defect detection algorithm based on deep learning. *Comput. Intell. Neurosci.* **2021**, *2021*, 1–13. [CrossRef]
34. Zhao, C.; Shu, X.; Yan, X.; Zuo, X.; Zhu, F. RDD-YOLO: A modified YOLO for detection of steel surface defects. *Measurement* **2023**, *214*, 112776. [CrossRef]
35. Wang, L.; Liu, X.; Ma, J.; Su, W.; Li, H. Real-Time Steel Surface Defect Detection with Improved Multi-Scale YOLO-v5. *Processes* **2023**, *11*, 1357. [CrossRef]
36. Li, M.; Wang, H.; Wan, Z. Surface defect detection of steel strips based on improved YOLOv4. *Comput. Electr. Eng.* **2022**, *102*, 108208. [CrossRef]
37. Liu, G.H.; Chu, M.X.; Gong, R.F.; Zheng, Z.H. DLF-YOLOF: An improved YOLOF-based surface defect detection for steel plate. *J. Iron Steel Res. Int.* **2023**, 1–10. [CrossRef]
38. Wang, R.-J.; Liang, F.-L.; Mou, X.-W.; Chen, L.-T.; Yu, X.-Y.; Peng, Z.-J.; Chen, H.-Y. Development of an Improved YOLOv7-Based Model for Detecting Defects on Strip Steel Surfaces. *Coatings* **2023**, *13*, 536. [CrossRef]
39. Shao, Y.; Fan, S.; Sun, H.; Tan, Z.; Cai, Y.; Zhang, C.; Zhang, L. Multi-Scale Lightweight Neural Network for Steel Surface Defect Detection. *Coatings* **2023**, *13*, 1202. [CrossRef]
40. Li, X.; Wang, W.; Wu, L.; Chen, S.; Hu, X.; Li, J.; Tang, J.; Yang, J. Generalized Focal Loss: Learning Qualified and Distributed Bounding Boxes for Dense Object Detection. *arXiv* **2020**, arXiv:2006.04388.
41. Zheng, Z.; Wang, P.; Liu, W.; Li, J.; Ye, R.; Ren, D. Distance-IoU loss: Faster and better learning for bounding box regression. In Proceedings of the AAAI Conference on Artificial Intelligence, New York, NY, USA, 7–12 February 2020; pp. 12993–13000.
42. Tong, Z.; Chen, Y.; Xu, Z.; Yu, R. Wise-IoU: Bounding Box Regression Loss with Dynamic Focusing Mechanism. *arXiv* **2023**, arXiv:2301.10051.
43. Chen, J.; Kao, S.-H.; He, H.; Zhuo, W.; Wen, S.; Lee, C.-H.; Chan, S.-H.G. Run, Don't walk: Chasing higher FLOPs for faster neural networks. In Proceedings of the IEEE/CVF Conference on Computer Vision and Pattern Recognition, Vancouver, BC, Canada, 18–22 June 2023.
44. Xu, X.; Jiang, Y.; Chen, W.; Huang, Y.; Zhang, Y.; Sun, X. DAMO-YOLO: A Report on Real-Time Object Detection Design. *arXiv* **2022**, arXiv:2211.15444.
45. Woo, S.; Park, J.; Lee, J.Y.; Kweon, I.S. Cbam: Convolutional block attention module. In Proceedings of the European Conference on Computer Vision, Munich, Germany, 8–14 September 2018; pp. 3–19.
46. Hu, J.; Shen, L.; Sun, G. Squeeze-and-excitation networks. In Proceedings of the IEEE Conference On Computer Vision and Pattern Recognition, New York, NY, USA, 18–23 June 2018; pp. 7132–7141.
47. Wang, Q.; Wu, B.; Zhu, P.; Li, P.; Zuo, W.; Hu, Q. ECA-Net: Efficient Channel Attention for Deep Convolutional Neural Networks. In Proceedings of the IEEE/CVF Conference on Computer Vision and Pattern Recognition, Seattle, WA, USA, 13–19 June 2020; pp. 11531–11539.
48. Hou, Q.; Zhou, D.; Feng, J. Coordinate attention for efficient mobile network design. In Proceedings of the IEEE/CVF Conference on Computer Vision and Pattern Recognition, Nashville, TN, USA, 20–25 June 2021; pp. 13713–13722.
49. Zhang, Q.L.; Yang, Y.B. Sa-net: Shuffle attention for deep convolutional neural networks. In Proceedings of the IEEE International Conference on Acoustics, Speech and Signal Processing, Toronto, ON, Canada, 6–11 June 2021; pp. 2235–2239.
50. Li, X.; Zhong, Z.; Wu, J.; Yang, Y.; Lin, Z.; Liu, H. Expectation-maximization attention networks for semantic segmentation. In Proceedings of the IEEE/CVF International Conference on Computer Vision, Seoul, Republic of Korea, 27 October–2 November 2019; pp. 9167–9176.
51. Available online: http://faculty.neu.edu.cn/songkechen/zh_CN/zdylm/263270/list/index.html (accessed on 12 December 2022).

52. Lv, X.; Duan, F.; Jiang, J.-J.; Fu, X.; Gan, L. Deep Metallic Surface Defect Detection: The New Benchmark and Detection Network. *Sensors* **2020**, *20*, 1562. [CrossRef]
53. Available online: https://tianchi.aliyun.com/competition/entrance/231682/information (accessed on 15 August 2018).
54. Huang, W.; Wei, P. A PCB Dataset for Defects Detection and Classification. *arXiv* **2019**, arXiv:1901.08204.

Disclaimer/Publisher's Note: The statements, opinions and data contained in all publications are solely those of the individual author(s) and contributor(s) and not of MDPI and/or the editor(s). MDPI and/or the editor(s) disclaim responsibility for any injury to people or property resulting from any ideas, methods, instructions or products referred to in the content.

Article

Low-Resolution Steel Surface Defects Classification Network Based on Autocorrelation Semantic Enhancement

Xiaoe Guo [1], Ke Gong [2] and Chunyue Lu [2,*]

[1] Techanical and Electrical Engineering Department, Shanxi Institute of Energy, Taiyuan 030600, China; guoxe@sxie.edu.cn
[2] College of Mechanical Engineering, North University of China, Taiyuan 030000, China; s202102046@st.nuc.edu.cn
* Correspondence: luchunyue@nuc.edu.cn

Abstract: Aiming at the problems of low-resolution steel surface defects imaging, such as defect type confusion, feature blurring, and low classification accuracy, this paper proposes an autocorrelation semantic enhancement network (ASENet) for the classification of steel surface defects. It mainly consists of a backbone network and an autocorrelation semantic enhancement module (ASE), in which the autocorrelation semantic enhancement module consists of three main learnable modules: the CS attention module, the autocorrelation computation module, and the contextual feature awareness module. Specifically, we first use the backbone network to extract the basic features of the image and then use the designed CS attention module to enhance the basic features. In addition, to capture different aspects of semantic objects, we use the autocorrelation module to compute the correlation between neighborhoods and contextualize the basic and augmented features to enhance the recognizability of the features. Experimental results show that our method produces significant results, and the classification accuracy reaches 96.24% on the NEU-CLS-64 dataset. Compared with ViT-B/16, Swin_t, ResNet50, Mobilenet_v3_small, Densenet121, Efficientnet_b2, and baseline, the accuracy is 9.43%, 5.15%, 4.87%, 3.34%, 3.28%, 3.01%, and 2.72% higher, respectively.

Keywords: surface defects; convolutional neural network; autocorrelation enhancement; attention mechanism

Citation: Guo, X.; Gong, K.; Lu, C. Low-Resolution Steel Surface Defects Classification Network Based on Autocorrelation Semantic Enhancement. *Coatings* **2023**, *13*, 2015. https://doi.org/10.3390/coatings13122015

Academic Editor: George A. Stanciu

Received: 18 September 2023
Revised: 23 November 2023
Accepted: 27 November 2023
Published: 28 November 2023

Copyright: © 2023 by the authors. Licensee MDPI, Basel, Switzerland. This article is an open access article distributed under the terms and conditions of the Creative Commons Attribution (CC BY) license (https://creativecommons.org/licenses/by/4.0/).

1. Introduction

In industrial production, steel is one of the basic materials for the manufacture of various mechanical equipment and components, and it plays an irreplaceable role in the aerospace industry, automobile manufacturing, shipbuilding, construction, the energy industry, and other fields [1]. However, the steel production process involves the coordinated operation of multiple pieces of equipment and complex procedures. If the equipment parameters are not set correctly or fail, it may lead to defects on the steel surface, such as oxidized skin, plaques, cracks, pitting, inclusions, scratches, and so on. These defects not only affect the appearance of steel but may also indicate that the steel has been damaged internally, seriously affecting the mechanical properties and corrosion resistance, which leads to a decline in product quality and even causes safety accidents [2]. Therefore, timely detection of steel surface defects and real-time adjustment of production equipment is essential to ensure steel quality and reduce production losses.

In the past, manual inspection sufficed to meet the demands of product output at relatively slower production speeds. However, with the escalation of production levels and the burgeoning market, manual review has become sluggish, inefficient, and labor-intensive. Prolonged hours of labor increase the chances of worker misdetection, while defect identification heavily relies on the inspector's experience. Furthermore, disparities exist in each worker's detection standards, making it arduous to fulfill production requirements.

Developing an accurate automatic detection solution using machine vision technology becomes imperative to overcome these challenges. Machine vision offers a non-contact and automated solution for surface defect detection [3–5]. However, it faces limitations in complex industrial environments, such as limited equipment universality, light-source requirements, and high costs. This hampers the efficiency of detection tasks.

With the development of computer vision technology, convolutional-neural-network-based detection has achieved wide application [6–9]. However, deep learning architectures have become larger and larger, resulting in an increasing number of parameters and requiring a large amount of computational resources [10,11]. Moreover, due to the fuzzy imaging and low resolution of steel defect images, the features learned by the network will suffer from information loss, feature blurring, as well as the problem of easy confusion. For this reason, ASENet, an autocorrelation-based semantic enhancement network for steel surface defects classification, is proposed, which mainly consists of a backbone network and an ASE module. Specifically, we first extract the basic features of the image through the backbone network, then use the CS attention module to enhance the basic features, and then use the autocorrelation module to compute the correlation between neighbors. Finally, we connect the enhanced features with the base features by residual concatenation to obtain self-attention features to capture different aspects of the semantic object. Experimental results show that our approach achieves state-of-the-art results on NEU-CLS-64 [12].

Our contributions are as follows:

- This paper proposes a new autocorrelation semantic enhancement method (ASE), which enhances the base features and extracts important local area features through CS attention and autocorrelation modules.
- By combining the backbone network and the autocorrelation semantic enhancement module, our model ASENet can solve the problems of information loss, feature ambiguity, and confusion that traditional neural network models would have when dealing with low-resolution steel defect images.
- Significant classification accuracies are achieved on the NEU-CLS-64 and CIFAR-100 datasets, and comparisons with several benchmark models demonstrate the effectiveness and superiority of the method.

2. Related Work

2.1. Classification of Steel Defects

Conventional methods for identifying defects on steel surfaces mainly use the wavelet transform [13,14] double-threshold binarization [15,16], and decision trees [17,18] to analyze and detect images. In addition to this, Mukhopadhyay et al. used multi-scale morphological segmentation of gray-scale images to process surface image data [19]; Podulka et al. used surface topographic image (STI) processing to characterize selected features from the surface texture [20]; and Ravimal et al. used the intensity of the near-field contrast image after reflecting light and reflective mirrors, as well as photometric stereoscopic techniques to recover the normal of the surface mapping for automated surface inspection [21]. By adopting these techniques, great strides have been made in optimizing productivity and improving product quality. However, these traditional methods have limitations. An obvious disadvantage is the relatively slow detection speed and the limited applicability. The algorithms used in these technologies often require substantial redesign for different application scenarios. This requirement not only increases the complexity of the process but also poses a significant obstacle to its widespread application. Another limiting factor is the high image quality requirement. When the input image is of low resolution, due to the small number of pixels in the low-resolution image, many details and information cannot be expressed in the picture, resulting in a decrease in detection accuracy.

Some recent approaches use convolutional neural networks for steel defect detection. These techniques are particularly effective in natural scenes, where they have gained substantial research support. For instance, Li et al. utilized coordinate attention and self-interaction to identify hot-rolled strip surface defects [22] effectively. Furthermore, Hao et al. proposed a novel two-stream neural network with sample generation and transfer learning for classifying steel strip surface defects [23]. Zhang et al. contributed to this growing research by proposing a novel approach for accurately classifying strip surface defects using generative adversarial networks and attention mechanisms [24]. Li et al. took a different approach by submitting a hybrid network architecture (CNN-T) that merged the CNN and Transformer encoders, achieving significant results on the NEU-CLS dataset [25]. However, these methods also require high-quality defect images.

Therefore, we proposed the ASENet network for low-resolution steel surface defect images to solve the problems of blurred edges and contours of low-resolution defect images, which lead to information loss, blurred features, and easy confusion.

2.2. Attention Mechanism

The purpose of the attention mechanism is to let the system learn to focus on areas of interest or high value from a large amount of information. It has now been successfully applied to various tasks [26–29]. For example, the self-attention mechanism used in the Transform model in 2017 has become a significant turning point in developing large-scale models [10]. Furthermore, SENet [30] introduces a channel attention block for image classification; it assigns attention weights to different input channels, allowing the network to focus on the most informative channels. Building upon this, the ECANet [31] further improved upon the SENet's strategy by proposing an adequate channel attention (ECA) block for convolutional neural networks (CNNs), successfully enabling cross-channel interaction. CBAM introduces the channel and spatial attention modules to establish the dual mechanism of channel and spatial attention [32]. While computing image autocorrelation, conventional approaches often rely on neighborhood correlation [33,34]. However, this approach is computationally intensive and adds complexity to the model.

In contrast, the ASE module proposed in this paper achieves contextual feature awareness through simple connections after enhancing basic features, eliminating the need for many redundant parameters. This module allows the model to capture dependencies between adjacent elements in the input, enhancing contextual understanding.

3. Our Approach

In this section, we focus on the specific implementation of ASENet. Figure 1 shows the overall architecture of the network. ASENet mainly consists of a backbone network and ASE modules. The ASE modules consist of three main learnable modules: the CS attention module, the autocorrelation computation module, and the contextual feature-aware module, as shown in Figure 2.

3.1. Overall Architecture

Given a set of steel defect sample images, we extract basic features $Z \in \mathbb{R}^{H \times W \times C}$ using a backbone network. Subsequently, we augment these basic features by the CS attention module to obtain the augmented feature representation $F \in \mathbb{R}^{H \times W \times C}$. The autocorrelation is then computed on the augmented feature F to obtain the autocorrelation tensor $Din \mathbb{R}^{H \times W \times C_1}$ ($C_1 = U \times V \times C$) and contextualized with the base feature Z to produce the self-attention feature $A \in \mathbb{R}^{H \times W \times C_g}$ ($C_g = C' + C$). Finally, the resulting features are passed through two output convolutional layers to recover the number of channels and residual to derive the final output $G \in \mathbb{R}^{H \times W \times C}$ from the basic features.

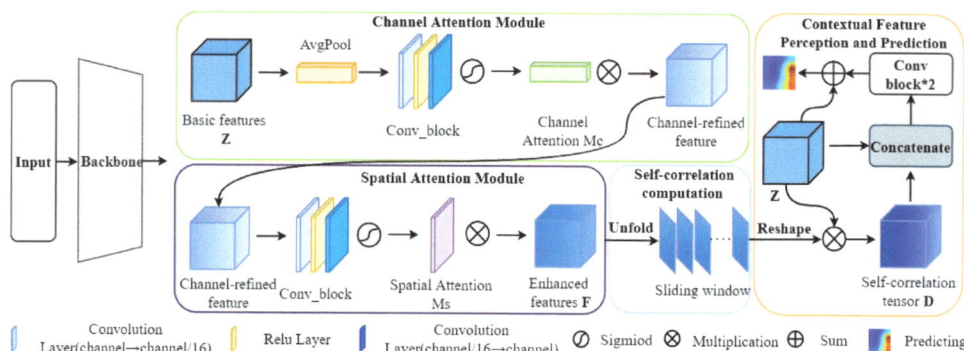

Figure 1. The overall architecture of ASENet.

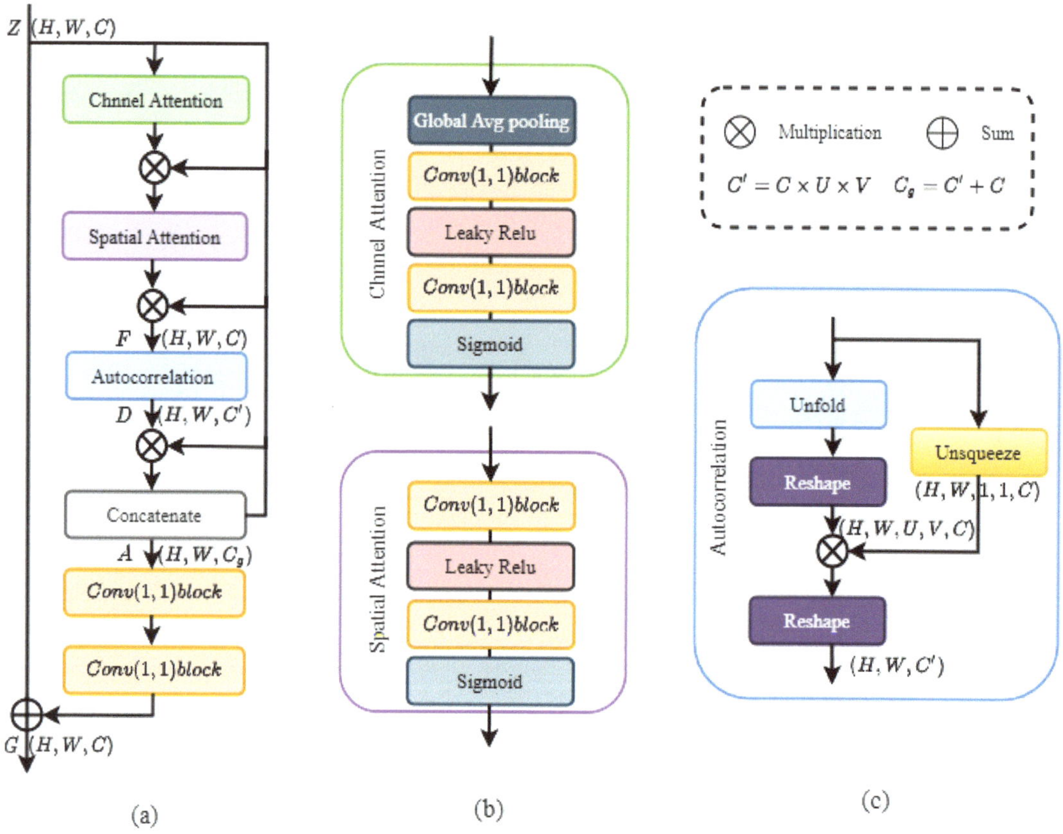

Figure 2. (a) ASE modules, (b) CS attention module, and (c) autocorrelation calculation.

3.2. CS Attention Module

We devised a CS attention method to enhance the channel and spatial information in the basic features. This method consists of two complete modules: the channel and spatial attention modules, as shown in Figure 2b. These modules collaborate to generate the attention weights, which are essential for determining the importance of each track and its spatial location in the feature.

We calculate each channel's attention weight within the channel attention module, denoted as M_c. Specifically, we begin by processing the foundational features using average pooling, followed by their refinement through a convolutional block. We subsequently map the final result to an output range between 0 and 1, achieved via a Sigmoid function, thereby obtaining the channel attention weight M_c. This weight effectively mirrors the importance of each channel in capturing pertinent information. We can prioritize the channels with information-rich content by assigning higher weights to channels housing valuable insights and lower weights to those contributing less.

Similarly, within the spatial attention module, we generate attention weights, denoted as M_s, for each spatial location within the feature. In this case, we employ a convolutional block to directly learn the features and subsequently map the results to an output range constrained between 0 and 1, once again utilizing a Sigmoid function. This process yields the spatial attention weight M_s, which signifies the significance of each pixel location in capturing meaningful insights. We can effectively concentrate on the pivotal spatial positions that drive understanding and context by attributing higher weights to areas that substantially contribute to the overall comprehension of the data and lower consequences to those of lesser informative value. Through the weights obtained by M_c and M_s, we can get the final enhanced feature $F \in \mathbb{R}^{H \times W \times C}$.

$$F = Z \times M_c \times M_s \qquad (1)$$

Unlike previous approaches [32], we have not utilized global max pooling to capture weights in channel and spatial dimensions or employed MLP (multi-layer perceptron) for feature extraction. Instead, we have used a convolutional approach to extract feature information. Additionally, in our CS attention module, we employed a single branch to learn the relationship between channel and spatial positions. Our experiments have revealed that lightweight attention modules are more suitable for low-resolution and blurry image processing tasks while avoiding unnecessary computations.

3.3. Autocorrelation Computation Calculation

To capture the self-similarity of neighborhoods in the image, we employ a calculation method that involves the Hadamard product. This product is performed between the C-dimensional vectors at each position x in the enhanced feature $F \in \mathbb{R}^{H \times W \times C}$ and their corresponding values in the neighborhood. The resulting products are then collected into a self-correlation tensor denoted as $D \in \mathbb{R}^{H \times W \times C_1}$. The self-correlation tensor D represents the relationships between different positions in the image. It allows us to identify patterns and similarities within local neighborhoods. We can obtain a tensor that expresses these relationships by calculating the Hadamard product and accumulating the results into D. The dimensionality of D is represented by C_1, corresponding to the channel of output vectors.

$$D(x,p) = \frac{Z(x)}{\|Z(x)\|} \odot \frac{F(x+p)}{\|F(x+p)\|} \qquad (2)$$

where $p \in [-d_U, d_U] \times [-d_V, d_V]$ corresponds to the relative positions in the neighborhood window. Here, d_U and d_V represent the maximum displacement in the horizontal and vertical directions, respectively. The window size is determined by $U = 2d_U + 1$ and $V = 2d_V + 1$, which includes the center position. Our experiments use a sliding window of $(U, V) \in (1, 1)$. Unlike previous work [35], we do not keep the dimension of $U \times V$, but consider it as part of the channel features so that we can obtain the new channel dimension $C_1 = U \times V \times C'$.

3.4. Contextual Feature Perception

Even though autocorrelation computing may determine how similar two images are to one another, it lacks the local semantic cues that the original convolutional features provided. We use a straightforward fusion step to create a contextual feature-aware semantic representation to capture various properties of the semantic objects. We specifically sew Z and D together to make the contextual semantic characteristics $A \in \mathbb{R}^{H \times W \times C_g}$, as shown below.

$$A_{(i,j)} = \left[Z_{(i,j)}^T, D_{(i,j)}^T \right]^T \quad (3)$$

To analyze the contextual relationships in A, convolution and bulk normalization operations are performed through the first convolutional layer to extract more semantically meaningful feature information. The extracted feature tensor is then subjected to a re-convolution operation to reduce the number of feature channels to the number of input channels to obtain a more compact feature representation. The convolution kernels for the two above convolutions' blocks are both of size 1×1. The convolution block $h()$ learns contextual relationships without padding and aggregates local correlation patterns, restoring the channel dimension to C, ensuring that the output $h(h(A))$ is the same size as Z. We then combine these two representations to generate the final feature representation $G \in \mathbb{R}^{H \times W \times C}$.

$$G = h(h(A)) + Z \quad (4)$$

Using an autocorrelation semantic enhancement network to augment the basic features helps to locate the essential regions of the target object. It enhances the recognizability of the features, which better achieves accurate classification in low-resolution images and improves the network's performance.

4. Experimental Results

In this section, we evaluate the performance of ASENet on the NEU-CLS-64 and CIFAR-100 datasets and compare it to other methods. In addition, we perform ablation studies and compare them with other attention methods to validate the effectiveness of the autocorrelation semantic enhancement networks. Figure 3 shows the test results of different models running 100 epochs on the NEU-CLS-64 dataset under the same experimental setup. Figure 4 shows the loss curves for the other models.

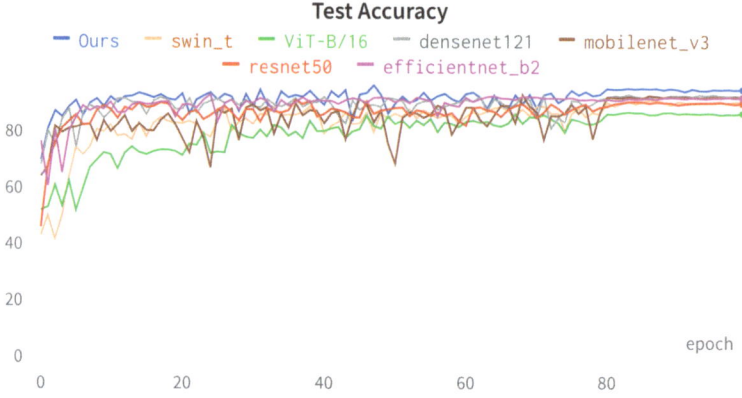

Figure 3. Test results of different models running 100 epochs on the NEU-CLS-64 dataset.

Figure 4. Loss curves of different models running on the NEU-CLS-64 dataset for 100 epochs.

4.1. Dataset

In the Northeastern University (NEU) Surface Defect Database [12], six typical surface defects of hot-rolled steel strips are collected, namely rolling scale (RS), patches (Pa), cracks (Cr), pitting surface (PS), inclusions (In), and scratches (Sc). The NEU-CLS-64 dataset used in this experiment includes an additional three defects: oil stains (Sp), pits (Gg), and rust (Rp). Furthermore, compared to the NEU dataset, all images in the NEU-CLS-64 dataset are of a fixed size of 64 × 64 pixels. The number of images per category varies (for instance, the pits (Gg) category has 296 images, while the oil stains (Sp) category has 438 images), as shown in Figure 5. CIFAR-100 is a widely used dataset in computer vision research. It consists of 60,000 color images, each of size 32 × 32 pixels, belonging to 100 different classes. The dataset is divided into two sets: a training set with 50,000 images and a test set with 10,000 images. This lower-resolution imagery undoubtedly poses more significant challenges for accurate classification by the neural network.

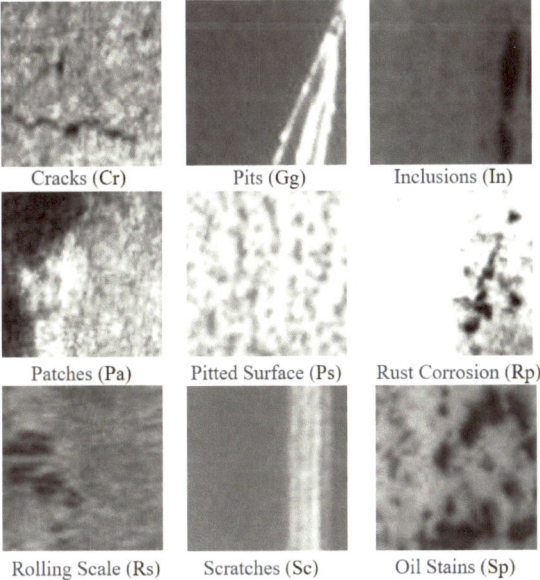

Figure 5. Sample images of 9 typical surface defects in the NEU-CLS-64 dataset.

4.2. Experimental Details

This study conducted experiments using the PyTorch 1.8 deep learning framework on a system equipped with an NVIDIA 2080Ti GPU and an Intel i7 9700K CPU. The NEU-CLS-64 dataset is divided into training and test sets in the ratio of 80:20, where 80% of the data is used for training, and the remaining 20% is used for testing, and CIFAR-100 uses 500 images of each class as the training set and 100 images as the test set. In these experiments, the NEU-CLS-64 input image size is 64×64 pixels, CIFAR-100 input image size is 32×32 pixels, ConvNet-4 is used for the backbone network, and 3×3 convolution kernels are used for each convolutional block. The number of channels in each Conv block increased to 64-160-320-640 to capture more semantic information in the feature maps. For optimization, the SGD (stochastic gradient descent) optimizer was used with a momentum of 0.9. The initial learning rate was set at 0.01, with a decay factor of 0.05. The training was performed on the NEU-CLS-64 dataset for 100 epochs, with a batch size of 64 samples. Learning rate annealing was applied after the 80th and 90th epochs by reducing the learning rate by 0.1.

4.3. Results

Table 1 shows that on the NEU-CLS-64 dataset, the proposed method outperforms all other methods in terms of accuracy, achieving an impressive 96.24%. Compared to ViT-B/16, Swin-t, ResNet50, MobileNetV3 Small, DenseNet121, and EfficientNetB2, it achieves an improvement of 9.43%, 5.15%, 4.87%, 3.34%, 3.28%, and 3.01%, respectively. Figure 3 clearly illustrates that ASENet consistently maintains a high correct rate throughout the testing process. In contrast, ViT-B/16 exhibits a relatively smoother but substantially lower correct rate compared to the other models. Turning our attention to the loss curve, it becomes evident that ASENet achieves the lowest loss among the seven different model types and exhibits the smoothest loss curve, as depicted in Figure 4. Our method also achieved the best results on the CIFAR-100 dataset, showing that ASENet performs better even on a 32×32 pixel dataset. When considering the parameter size, our method stands out with only 3.04 MB, which is significantly smaller compared to other approaches. This implies that our method is more efficient in terms of memory usage.

Table 1. Comparison results with other methods on the NEU-CLS-64 and CIFAR-100 dataset.

Method	NEU-CLS-64	CIFAR-100	Params Size (M)	FLOPs (G)
ViT-B/16 [36]	86.81	56.19	326.74 MB	93.09 GFLOPs
Swin_t [37]	91.09	58.24	71.94 MB	47.33 GFLOPs
ResNet50 [38]	91.37	61.08	89.75 MB	21.59 GFLOPs
Mobilenet_v3_small [39]	92.90	59.44	5.83 MB	0.38 GFLOPs
Densenet121 [40]	92.96	66.30	31.01 MB	15.13 GFLOPs
Efficientnet_b2 [41]	93.23	60.32	34.75 MB	3.74 GFLOPs
ASENet (Ours)	**96.24**	**71.66**	**3.04 MB**	17.61 GFLOPs

It is worth noting that the latest ViT and Swin models exhibited the worst performance during the experiments. These two models have the most complex computations and parameters and deliver the poorest results. This can be attributed to the fact that when the input image size is small, the divided sub-patches are relatively tiny, containing limited information in each sub-patch. This limitation hinders the classifier from capturing sufficient information, thus affecting the model's performance. Additionally, Transformer encoders have difficulty handling local features effectively. Therefore, traditional convolutional neural network (CNN) models perform better when dealing with low-resolution images.

4.4. Ablation Studies

To evaluate the effectiveness of ASENet, we created a baseline model (ConvNet-4) without any additional modules. We performed ablation experiments on the NEU-CLS-64 dataset. As shown in Table 2, we compare the three learnable modules in two-by-two combinations with the baseline model and ASENet. In scenario (b), where only autocorrelation computation and contextual features are used, the accuracy achieved is 95.30%. Adding the CS attention module in method (c) slightly increases the accuracy to 95.47%. Similarly, including autocorrelation computation in scenario (d) with the CS attention module but without contextual features results in an accuracy of 95.68%. However, the most impressive improvement in accuracy is observed in scenario (e), where all three modules are used concurrently. An accuracy of 96.24% is achieved, the highest among all the methods.

Moreover, to analyze the impact of different values of the sliding window (U, V) on the network's autocorrelation computation, we conducted a comparison between $(U, V) \in (1,1)$, $(U, V) \in (3,3)$, and $(U, V) \in (5,5)$, which we named as A, B, and C, respectively. It is important to note that $(U, V) \in (1,1)$, referred to as A, represents the sliding window used by ASENet. From Figure 6, it can be seen that as the value of (U, V) increases gradually, the network performance decreases, which is due to the small size of the defective image, though the backbone-network-extracted feature maps are also relatively small, so a larger sliding window will lead to the loss of important information in the extraction process.

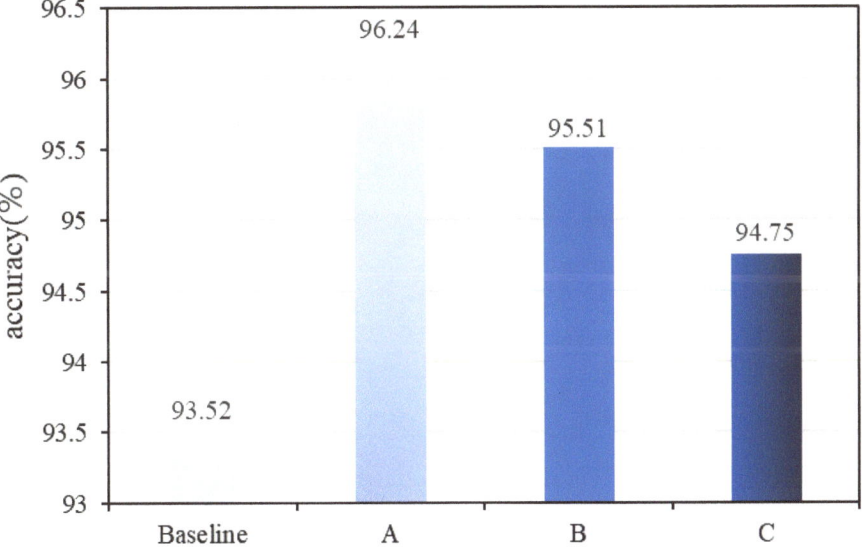

Figure 6. The impact of U and V values on the network.

Table 2. Adding the effects of different modules.

Id	CS Attention	Autocorrelation Computation	Contextual Feature	Accuracy (%)
(a)	×	×	×	93.52
(b)	×	✓	✓	95.30
(c)	✓	×	✓	95.47
(d)	✓	✓	×	95.68
(e)	✓	✓	✓	**96.24**

4.5. Comparison with Other Attention Modules

Table 3 compares the accuracy, number of parameters, and computational complexity of the autocorrelation semantic enhancement modules (ASE) and other attention modules. According to the study results, ASE outperforms other attention modules on NEU-CLS-64 by 2.72% over the baseline model. Although ASE performs well in accuracy, it is not the lightest in model parameter size. Regarding parameter size, ShuffleAttention is optimal at only 0.05 MB, while ASE has 1.27 MB of parameters. However, a slight parameter increase compared to performance may be acceptable, especially if computational resources allow it. Regarding computational complexity, our method is comparable mainly to the SE module, achieving the highest accuracy while maintaining competitive parameter sizes and computational complexity. This demonstrates the effectiveness of our approach in capturing important features and improving classification accuracy.

Table 3. Comparison results with other attention modules.

Method	Accuracy (%)	Params Size (M)	FLOPs (G)
Baseline	93.52	0 MB	0 GFLOPs
TripletAttention [42]	95.03	0.06 MB	0.14 GFLOPs
ShuffleAttention [43]	95.37	0.05 MB	0.01 GFLOPs
PSA [44]	95.44	0.78 MB	10.10 GFLOPs
CoTAttention [45]	95.45	1.08 MB	18.56 GFLOPs
MobileViTv2Attention [46]	95.51	1.17 MB	20.14 GFLOPs
Coord_attention [47]	95.65	0.07 MB	0.13 GFLOPs
SE [30]	95.68	0.08 MB	1.39 GFLOPs
CBAM [32]	95.76	0.10 MB	0.02 GFLOPs
ASE (Ours)	**96.24**	1.27 MB	1.32 GFLOPs

4.6. Visualisation

Figure 7 showcases heatmap visualizations of various models [48]. The illustration reveals that the attention patterns in both the ViT and Swin models exhibit a more dispersed distribution. Conversely, the approach advocated in this paper, enhanced by the autocorrelation semantic enhancement module, effectively sieves out extraneous regions, channeling attention towards more pivotal image characteristics. Particularly for defects such as Cr, In, Pa, Rp, and Rs, it adeptly pinpoints their locations with superior accuracy compared to alternative methods. Furthermore, ASENet amalgamates features across diverse scales, rendering it more adept at discerning valuable features in intricate settings.

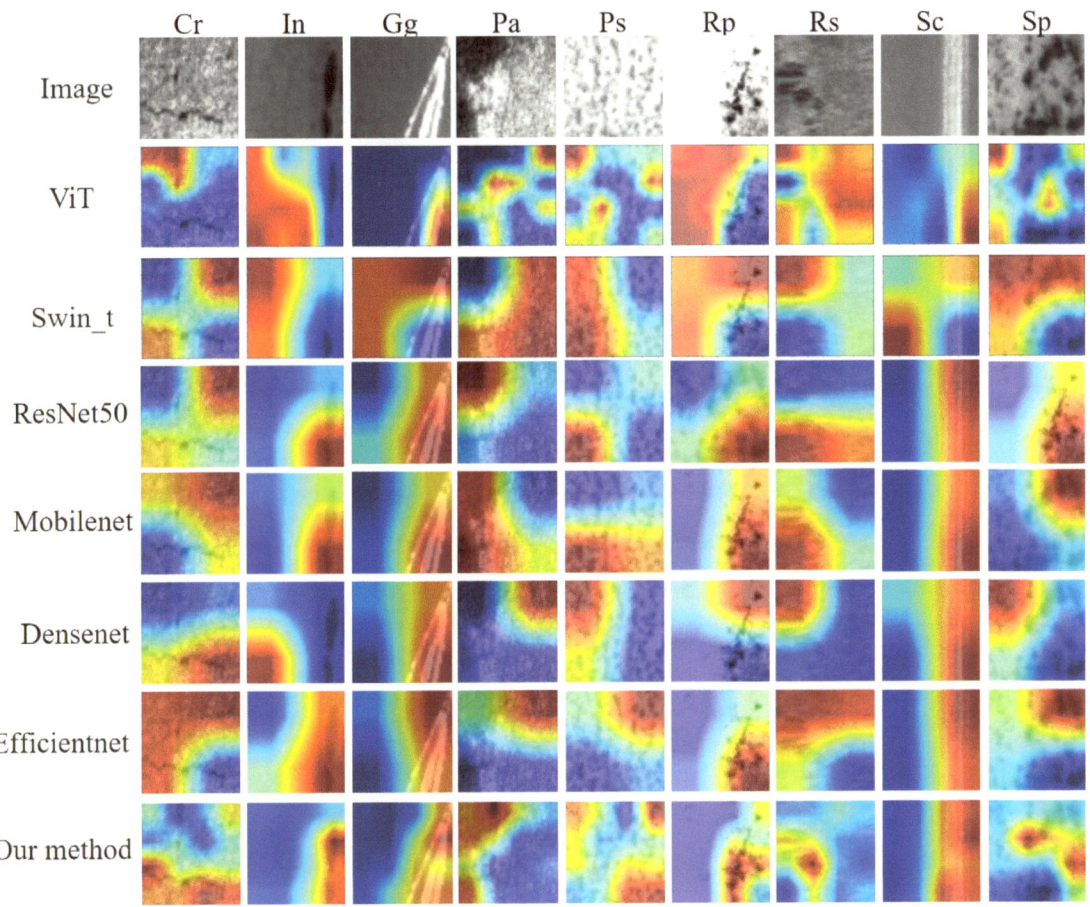

Figure 7. Results of different model visualizations.

5. Conclusions

In this study, we propose the autocorrelation semantic enhancement network (ASENet) to address the challenge of imaging defects on low-resolution steel surfaces. ASENet consists of a backbone network and an ASE module, where the ASE contains a CS attention module, an autocorrelation computation module, and a contextual feature-aware module. The ASE module captures different aspects of feature semantics and enhances feature recognizability by augmenting the underlying features and computing correlations between neighboring domains. Experimental results on NEU-CLS-64 datasets show that the proposed ASENet can effectively solve the problems of defect type confusion, feature ambiguity, and low classification accuracy in low-resolution steel surface defect imaging. Compared with the existing models, the enhanced feature recognition capability of ASENet gives it superior performance, making it a promising method for steel surface defect classification.

Author Contributions: X.G. played a significant role in the conceptualization and design of the study, as well as the acquisition and analysis of the data and the experimental design. K.G. made substantial contributions to data collection and analysis. He also contributed to the writing and revising of the manuscript, ensuring its overall coherence and clarity. C.L. contributed to the interpretation of the data, critically reviewed the manuscript for intellectual content, and provided valuable insights and suggestions for improvement throughout the research process. All authors have read and agreed to the published version of the manuscript.

Funding: This work was supported by the Shanxi Key RD Program, Project No. 201903D121063.

Institutional Review Board Statement: Not applicable.

Informed Consent Statement: Not applicable.

Data Availability Statement: Data is contained within the article.

Conflicts of Interest: The authors declare no conflict of interest.

References

1. Gardner, L. The use of stainless steel in structures. *Prog. Struct. Eng. Mater.* **2010**, *7*, 45–55. [CrossRef]
2. Wen, X.; Shan, J.; He, Y.; Song, K. Steel surface defect recognition: A survey. *Coatings* **2022**, *13*, 17. [CrossRef]
3. Chu, M.; Zhao, J.; Liu, X.; Gong, R. Multi-class classification for steel surface defects based on machine learning with quantile hyper-spheres. *Chemom. Intell. Lab. Syst.* **2017**, *168*, 15–27. [CrossRef]
4. Park, J.K.; Kwon, B.K.; Park, J.H.; Kang, D.J. Machine learning-based imaging system for surface defect inspection. *Int. J. Precis. Eng. -Manuf.-Green Technol.* **2016**, *3*, 303–310. [CrossRef]
5. Tang, B.; Chen, L.; Sun, W.; Lin, Z.k. Review of surface defect detection of steel products based on machine vision. *IET Image Process.* **2023**, *17*, 303–322. [CrossRef]
6. Wang, C.Y.; Bochkovskiy, A.; Liao, H.Y.M. YOLOv7: Trainable bag-of-freebies sets new state-of-the-art for real-time object detectors. In Proceedings of the IEEE/CVF Conference on Computer Vision and Pattern Recognition, Vancouver, BC, Canada, 8–22 June 2023; pp. 7464–7475. [CrossRef]
7. Cai, Z.; Vasconcelos, N. Cascade r-cnn: Delving into high quality object detection. In Proceedings of the IEEE Conference on Computer Vision and Pattern Recognition, Salt Lake City, UT, USA, 18–23 June 2018; pp. 6154–6162. [CrossRef]
8. Tian, Z.; Shen, C.; Chen, H.; He, T. Fcos: Fully convolutional one-stage object detection. In Proceedings of the IEEE/CVF International Conference on Computer Vision, Seoul, Republic of Korea, 27 October–2 November 2019; pp. 9627–9636.
9. Tan, M.; Pang, R.; Le, Q.V. Efficientdet: Scalable and efficient object detection. In Proceedings of the IEEE/CVF Conference on Computer Vision and Pattern Recognition, Seattle, WA, USA, 13–19 June 2020; pp. 10781–10790. [CrossRef]
10. Vaswani, A.; Shazeer, N.; Parmar, N.; Uszkoreit, J.; Jones, L.; Gomez, A.N.; Kaiser, Ł.; Polosukhin, I. Attention is all you need. *Adv. Neural Inf. Process. Syst.* **2017**, *30*.
11. Doersch, C. Tutorial on variational autoencoders. *arXiv* **2016**, arXiv:1606.05908. [CrossRef]
12. Song, K.; Yan, Y. A noise robust method based on completed local binary patterns for hot-rolled steel strip surface defects. *Appl. Surf. Sci.* **2013**, *285*, 858–864. [CrossRef]
13. Zhang, S.; Karim, M. A new impulse detector for switching median filters. *IEEE Signal Process. Lett.* **2002**, *9*, 360–363. [CrossRef]
14. Wu, X.y.; Xu, K.; Xu, J.w. Application of Undecimated Wavelet Transform to Surface Defect Detection of Hot Rolled Steel Plates. In Proceedings of the 2008 Congress on Image and Signal Processing, Sanya, China, 27–30 May 2008; Volume 4, pp. 528–532. [CrossRef]
15. Senthikumar, M.; Palanisamy, V.; Jaya, J. Metal surface defect detection using iterative thresholding technique. In Proceedings of the Second International Conference on Current Trends In Engineering and Technology—ICCTET 2014, Coimbatore, India, 8 July 2014; pp. 561–564. [CrossRef]
16. Yun, J.P.; Kim, D.; Kim, K.; Lee, S.J.; Park, C.H.; Kim, S.W. Vision-based surface defect inspection for thick steel plates. *Opt. Eng.* **2017**, *56*, 053108. [CrossRef]
17. Aghdam, S.R.; Amid, E.; Imani, M.F. A fast method of steel surface defect detection using decision trees applied to LBP based features. In Proceedings of the 2012 7th IEEE Conference on Industrial Electronics and Applications (ICIEA), Singapore, 18–20 July 2012; pp. 1447–1452. [CrossRef]
18. Jian, L.; Wei, H.; Bin, H. Research on inspection and classification of leather surface defects based on neural network and decision tree. In Proceedings of the 2010 International Conference On Computer Design and Applications, Qinhuangdao, China, 25–27 June 2010; Volume 2, pp. V2-381–V2-384. [CrossRef]
19. Mukhopadhyay, S.; Chanda, B. Multiscale morphological segmentation of gray-scale images. *IEEE Trans. Image Process.* **2003**, *12*, 533–549. [CrossRef] [PubMed]
20. Podulka, P. Application of image processing methods for the characterization of selected features and wear analysis in surface topography measurements. *Procedia Manuf.* **2021**, *53*, 136–147. [CrossRef]
21. Ravimal, D.; Kim, H.; Koh, D.; Hong, J.H.; Lee, S.K. Image-based inspection technique of a machined metal surface for an unmanned lapping process. *Int. J. Precis. Eng. Manuf. Green Technol.* **2020**, *7*, 547–557. [CrossRef]
22. Li, Z.; Wu, C.; Han, Q.; Hou, M.; Chen, G.; Weng, T. CASI-Net: A novel and effect steel surface defect classification method based on coordinate attention and self-interaction mechanism. *Mathematics* **2022**, *10*, 963. [CrossRef]

23. Hao, Z.; Li, Z.; Ren, F.; Lv, S.; Ni, H. Strip steel surface defects classification based on generative adversarial network and attention mechanism. *Metals* **2022**, *12*, 311. [CrossRef]
24. Zhang, J.; Li, S.; Yan, Y.; Ni, Z.; Ni, H. Surface Defect Classification of Steel Strip with Few Samples Based on Dual-Stream Neural Network. *Steel Res. Int.* **2022**, *93*, 2100554. [CrossRef]
25. Li, S.; Wu, C.; Xiong, N. Hybrid architecture based on CNN and transformer for strip steel surface defect classification. *Electronics* **2022**, *11*, 1200. [CrossRef]
26. Bello, I.; Zoph, B.; Vaswani, A.; Shlens, J.; Le, Q.V. Attention augmented convolutional networks. In Proceedings of the IEEE/CVF International Conference on Computer Vision, Seoul, Republic of Korea, 27 October–2 November 2019; pp. 3286–3295.
27. Hu, D. An introductory survey on attention mechanisms in NLP problems. In Proceedings of the Intelligent Systems and Applications: Proceedings of the 2019 Intelligent Systems Conference (IntelliSys), London, UK, 3–4 September 2020; Volume 2, pp. 432–448. [CrossRef]
28. Fukui, H.; Hirakawa, T.; Yamashita, T.; Fujiyoshi, H. Attention branch network: Learning of attention mechanism for visual explanation. In Proceedings of the IEEE/CVF Conference on Computer Vision and Pattern Recognition, Long Beach, CA, USA, 15–20 June 2019; pp. 10705–10714.
29. Guo, M.H.; Xu, T.X.; Liu, J.J.; Liu, Z.N.; Jiang, P.T.; Mu, T.J.; Zhang, S.H.; Martin, R.R.; Cheng, M.M.; Hu, S.M. Attention mechanisms in computer vision: A survey. *Comput. Vis. Media* **2022**, *8*, 331–368. [CrossRef]
30. Hu, J.; Shen, L.; Sun, G. Squeeze-and-excitation networks. In Proceedings of the IEEE Conference on Computer Vision and Pattern Recognition, Salt Lake City, UT, USA, 18–23 June 2018; pp. 7132–7141. [CrossRef]
31. Wang, Q.; Wu, B.; Zhu, P.; Li, P.; Zuo, W.; Hu, Q. ECA-Net: Efficient channel attention for deep convolutional neural networks. In Proceedings of the IEEE/CVF Conference on Computer Vision and Pattern Recognition, Seattle, WA, USA, 13–19 June 2020; pp. 11534–11542.
32. Woo, S.; Park, J.; Lee, J.Y.; Kweon, I.S. Cbam: Convolutional block attention module. In Proceedings of the European Conference on Computer Vision (ECCV), Munich, Germany, 8–14 September 2018; pp. 3–19. [CrossRef]
33. Rocco, I.; Cimpoi, M.; Arandjelović, R.; Torii, A.; Pajdla, T.; Sivic, J. Neighbourhood consensus networks. *Adv. Neural Inf. Process. Syst.* **2018**, *31*.
34. Hassani, A.; Walton, S.; Li, J.; Li, S.; Shi, H. Neighborhood attention transformer. In Proceedings of the IEEE/CVF Conference on Computer Vision and Pattern Recognition, Vancouver, BC, Canada, 18–22 June 2023; pp. 6185–6194. [CrossRef]
35. Kang, D.; Kwon, H.; Min, J.; Cho, M. Relational embedding for few-shot classification. In Proceedings of the IEEE/CVF International Conference on Computer Vision, Montreal, QC, Canada, 10–17 October 2021; pp. 8822–8833. [CrossRef]
36. Dosovitskiy, A.; Beyer, L.; Kolesnikov, A.; Weissenborn, D.; Zhai, X.; Unterthiner, T.; Dehghani, M.; Minderer, M.; Heigold, G.; Gelly, S.; et al. An image is worth 16x16 words: Transformers for image recognition at scale. *arXiv* **2020**, arXiv:2010.11929. [CrossRef]
37. Liu, Z.; Lin, Y.; Cao, Y.; Hu, H.; Wei, Y.; Zhang, Z.; Lin, S.; Guo, B. Swin transformer: Hierarchical vision transformer using shifted windows. In Proceedings of the IEEE/CVF International Conference on Computer Vision, Montreal, BC, Canada, 11–17 October 2021; pp. 10012–10022. [CrossRef]
38. He, K.; Zhang, X.; Ren, S.; Sun, J. Deep residual learning for image recognition. In Proceedings of the IEEE Conference on Computer Vision and Pattern Recognition, Las Vegas, NV, USA, 27–30 June 2016; pp. 770–778.
39. Howard, A.; Sandler, M.; Chu, G.; Chen, L.C.; Chen, B.; Tan, M.; Wang, W.; Zhu, Y.; Pang, R.; Vasudevan, V.; et al. Searching for mobilenetv3. In Proceedings of the IEEE/CVF International Conference on Computer Vision, Seoul, Republic of Korea, 27 October–2 November 2019; pp. 1314–1324.
40. Huang, G.; Liu, Z.; Van Der Maaten, L.; Weinberger, K.Q. Densely connected convolutional networks. In Proceedings of the IEEE Conference on Computer Vision and Pattern Recognition, Honolulu, HI, USA, 21–26 July 2017; pp. 4700–4708. [CrossRef]
41. Tan, M.; Le, Q. Efficientnet: Rethinking model scaling for convolutional neural networks. In Proceedings of the International Conference on Machine Learning. PMLR, Long Beach, CA, USA, 9–15 June 2019; pp. 6105–6114.
42. Misra, D.; Nalamada, T.; Arasanipalai, A.U.; Hou, Q. Rotate to attend: Convolutional triplet attention module. In Proceedings of the IEEE/CVF Winter Conference on Applications of Computer Vision, Virtual Conference, 5–9 January 2021; pp. 3139–3148. [CrossRef]
43. Zhang, Q.L.; Yang, Y.B. SA-Net: Shuffle Attention for Deep Convolutional Neural Networks. In Proceedings of the ICASSP 2021—2021 IEEE International Conference on Acoustics, Speech and Signal Processing (ICASSP), Toronto, ON, Canada, 6–11 June 2021; pp. 2235–2239. [CrossRef]
44. Liu, H.; Liu, F.; Fan, X.; Huang, D. Polarized self-attention: Towards high-quality pixel-wise regression. *arXiv* **2021**, arXiv:2107.00782. [CrossRef]
45. Dai, Z.; Liu, H.; Le, Q.V.; Tan, M. Coatnet: Marrying convolution and attention for all data sizes. *Adv. Neural Inf. Process. Syst.* **2021**, *34*, 3965–3977.
46. Mehta, S.; Rastegari, M. Separable self-attention for mobile vision transformers. *arXiv* **2022**, arXiv:2206.02680. [CrossRef]

47. Hou, Q.; Zhou, D.; Feng, J. Coordinate attention for efficient mobile network design. In Proceedings of the IEEE/CVF Conference on Computer Vision and Pattern Recognition, Nashville, TN, USA, 20–25 June 2021; pp. 13713–13722. [CrossRef]
48. Selvaraju, R.R.; Cogswell, M.; Das, A.; Vedantam, R.; Parikh, D.; Batra, D. Grad-cam: Visual explanations from deep networks via gradient-based localization. In Proceedings of the IEEE International Conference on Computer Vision, Venice, Italy, 22–29 October 2017; pp. 618–626.

Disclaimer/Publisher's Note: The statements, opinions and data contained in all publications are solely those of the individual author(s) and contributor(s) and not of MDPI and/or the editor(s). MDPI and/or the editor(s) disclaim responsibility for any injury to people or property resulting from any ideas, methods, instructions or products referred to in the content.

MDPI AG
Grosspeteranlage 5
4052 Basel
Switzerland
Tel.: +41 61 683 77 34

Coatings Editorial Office
E-mail: coatings@mdpi.com
www.mdpi.com/journal/coatings

Disclaimer/Publisher's Note: The title and front matter of this reprint are at the discretion of the Guest Editor. The publisher is not responsible for their content or any associated concerns. The statements, opinions and data contained in all individual articles are solely those of the individual Editor and contributors and not of MDPI. MDPI disclaims responsibility for any injury to people or property resulting from any ideas, methods, instructions or products referred to in the content.

www.ingramcontent.com/pod-product-compliance
Lightning Source LLC
LaVergne TN
LVHW072354090526
838202LV00019B/2540